ROAD TRANSPORT RESEARCH

road transport research
– outlook 2000 –

THIRTIETH ANNIVERSARY

ORGANISATION FOR ECONOMIC CO-OPERATION AND DEVELOPMENT

ORGANISATION FOR ECONOMIC CO-OPERATION AND DEVELOPMENT

Pursuant to Article 1 of the Convention signed in Paris on 14th December 1960, and which came into force on 30th September 1961, the Organisation for Economic Co-operation and Development (OECD) shall promote policies designed:

- to achieve the highest sustainable economic growth and employment and a rising standard of living in Member countries, while maintaining financial stability, and thus to contribute to the development of the world economy;
- to contribute to sound economic expansion in Member as well as non-member countries in the process of economic development; and
- to contribute to the expansion of world trade on a multilateral, non-discriminatory basis in accordance with international obligations.

The original Member countries of the OECD are Austria, Belgium, Canada, Denmark, France, Germany, Greece, Iceland, Ireland, Italy, Luxembourg, the Netherlands, Norway, Portugal, Spain, Sweden, Switzerland, Turkey, the United Kingdom and the United States. The following countries became Members subsequently through accession at the dates indicated hereafter: Japan (28th April 1964), Finland (28th January 1969), Australia (7th June 1971), New Zealand (29th May 1973), Mexico (18th May 1994), the Czech Republic (21st December 1995), Hungary (7th May 1996), Poland (22nd November 1996) and the Republic of Korea (12th December 1996). The Commission of the European Communities takes part in the work of the OECD (Article 13 of the OECD Convention).

Publié en français sous le titre :

RECHERCHE EN MATIÈRE DE ROUTES ET DE TRANSPORTS ROUTIERS – PERSPECTIVES 2000

FOREWORD

The Programme centres on road and road transport research, while taking into account the impacts of intermodal aspects on the road transport system as a whole. It is geared towards a technico-economic approach to solving key road transport issues identified by Member countries. The Programme has two main fields of activity:

- International research and policy assessments of road and road transport issues to provide scientific support for decisions by Member governments and international governmental organisations;

- Technology transfer and information exchange through two databases - the International Road Research Documentation (IRRD) scheme and the International Road Traffic and Accident Database (IRTAD).

Its mission is to:

- Enhance innovative research through international co-operation and networking;
- Undertake joint policy analyses and prepare technology reviews of critical road transport issues;
- Promote the exchange of scientific and technical information in the transport sector and contribute to road technology transfer in OECD Member and non-member countries.

The scientific and technical activities concern:

- Infrastructure research;
- Road traffic and intermodal transport;
- Environment/transport interactions;
- Traffic safety research;
- Strategic research planning.

3

ABSTRACT

IRRD No. 887394

The report prepared for the 30th Anniversary of the OECD Road Transport Research Programme (RTR) highlights achievements from OECD activities during the period between 1992 and 1997. 40 international studies and other projects were undertaken during the period. These initiatives were carried out under the auspices of Scientific Expert Groups, Research Workshops, Seminars, Symposia, Conferences, and Joint Research Programmes. Chapter I is an introduction to the Programme highlighting the importance of international research co-operation for policy development and improvements of road technologies. Chapter II discusses Intermodality and Logistics and their relation to a growing global economy. Chapter III stresses advanced traffic technologies and management practices that help to improve road transport safety and efficiency. Chapter IV provides an overview of issues related to infrastructure maintenance and management. Chapter V describes the road safety studies undertaken and how research is helping to reduce traffic accident injuries and fatalities. Chapter VI discusses how transport-related activities impact the environment and explains how new approaches and practices are helping to minimise these impacts. Chapter VII describes the Programme's two international databases – IRRD and IRTAD – that allow sharing of information worldwide and dissemination. Chapter VIII defines the role for the Programme with non OECD member countries and a number of technology transfer activities focusing on maintenance issues and road safety initiatives. In total, the report provides an overview of research trends, results, and policy implications for a wide variety of topics and issues of concern to all 29 OECD Member countries.

Subject Classification: Traffic and transport planning; design of roads and related structures; environment; economics and administration.

Fields: 72, 20, 15, 10

Keywords: OECD; policy; research project; planning; freight transport; intermodal transport; intelligent transport system; safety; environment protection; data bank; developing countries; accident rate; technology; logistics; impact study (environment); accident prevention; international.

TABLE OF CONTENTS

EXECUTIVE SUMMARY

In December 1967, the OECD Council established the Road Research Programme. In the 30 years since that decision, Member countries have exhibited their interest through continuing involvement in the Programme, now called Road Transport Research, the participation of 40 Member and associated institutions in two operational international databases, and the active engagement of more than 3,000 scientific, engineering, and economic experts in the investigative activities. Over the years, the Programme has thus become a true knowledge bank for solutions to the challenges faced by road transport professionals.

MOTIVATION

The importance of roads and road transport to OECD economies is the raison d'être of the Programme. For all countries roads are a major national investment. Most OECD Member countries have mature road systems that require regular maintenance investments to ensure they remain useable for the travelling public. At the same time, the costs associated with road transport in terms of traffic accidents, the impact on the environment, and the use of non-renewable resources are significant. In this regard, all countries, but especially the OECD Member countries, share a common interest in finding the means to reduce the high costs associated with road transport. Road transport research is therefore an essential element in all national road programmes.

International co-operation in research has become much more of a strategic element in national research programmes. The majority of OECD countries have solid research programmes, up-to-date facilities and a sound scientific community. But policy makers increasingly ask if, on the one hand, the existing system can meet the needs of the next ten, twenty, or even thirty years while, on the other hand, emphasising the importance of short-term policy research priorities. Thus, the primary driving force for the Programme is the recognition by Member countries of the common challenges they share in the implementation of key transport and infrastructure policies and programmes.

The consensus on the common problems to investigate within the Programme avoids overlap of efforts, nationally and internationally, and assures the economic use of scarce resources. Joint fact-finding and research exchange minimise the risk of formulating ineffective or irrelevant conclusions and recommendations in Member countries. The Programme is a fundamental partner of the transport agencies in ensuring that road investments are made wisely, that economic contributions to transport are effective, and that road transport is as safe and environmentally sound as possible. It is in this context of major shared challenges that the Programme operates. As long as the OECD Member countries are faced with common problems it is in their benefit to work together to find solutions. The Programme provides an efficient mechanism for this co-operation.

THEMES

This 30th Anniversary report is designed to give an in-depth overview of the activities encompassed by the OECD Road Transport Research Programme. Additionally, readers will be left with a sound understanding of the continuing contributions of the Programme as well as the prospects it holds for bringing focused international attention to the pressing transport needs facing the OECD Member countries and all countries world-wide.

The Report is organised as follows:

POLICY RESEARCH – an overview of the Programme with discussion on its functions within the framework of OECD, its operating relationship with Member Countries, and the technical and policy interactions that drive and invigorate the Programme;

INTERMODALITY AND LOGISTICS – these concepts are critical to a growing global economy and will dramatically affect individual transport mode planning, operations and interaction in the years to come;

ADVANCED TRAFFIC TECHNOLOGIES AND PRACTICES – helping to improve the efficiency of road and transport systems by enhancing capacity through applications of new technologies and management approaches;

INFRASTRUCTURE – infrastructure design, construction, maintenance and management are the foundation of a good road programme and continue to be of major importance to all Member countries;

ROAD SAFETY – internationally co-ordinated research is helping to decrease the number of traffic accident injuries and fatalities and reduce accident costs which can amount to as much as 4 percent of GDP;

ENVIRONMENT/TRANSPORT INTERACTIONS – transport-related activities significantly impact the natural environment but new management and practices are helping to attenuate these impacts;

INTERNATIONAL DATABASES – two databases are maintained under the auspices of the Programme and allow extensive distribution and sharing of information that is important to all transport professionals; and

TECHNOLOGY TRANSFER – a role for the Programme with non-OECD Member countries has been defined and steps are being taken to share technology and best-practice techniques throughout the world.

EXPANDING HORIZONS

Member countries share a common commitment to exchange information with non-member countries. In the past three years, the Programme has supported Member countries in meeting their commitments with outreach efforts directed at co-operation with OECD non-member countries through technology transfer activities. The principal thrust of the activities undertaken is generally along the lines of providing technical or policy advice. Regional safety conferences in Asia, Latin America, and Africa are good examples of how the Programme works to bring state-of-the-practice information from Member to non-member countries to address a technical area that can have a profound impact on the targeted economies. In the coming years the Programme will be working with Member countries to build a more comprehensive management structure for the delivery of technology transfer initiatives.

A VISION FOR THE FUTURE

In the same manner that the OECD and the Steering Committee must look back to gauge results, so must they continually look ahead and consider what factors may impact transport policy and infrastructure investments. Through these considerations, the Programme can develop and design activities that will address both the immediate needs of the OECD Member countries as well as contribute to future challenges that will likely confront these same countries as transport demands change.

Among the significant trends that will affect the conduct and output of the Programme over the next triennial period is the requirement for an integrated intermodal systems approach assuring connectivity between road systems and other forms of transport. Advanced integrated logistics, backed up by up-to-date information technologies and communications networks, will be a key integrator in the planning and management process governing the transport sector. As the shift from individual modal approaches to an intermodal perspective continues, Member countries will be confronted by numerous challenges such as identifying critical or missing connections, developing new infrastructure, and addressing institutional shortcomings. The Programme is prepared to adopt a comprehensive multi-modal transport research approach which should also investigate economic incentives and barriers as well as innovative public/private financing mechanisms.

CHAPTER I: POLICY RESEARCH

I.1. SERVING THE NEEDS OF OECD COUNTRIES

In December 1967, the OECD Council, in its 132nd decision (see C(67)132/FINAL), established the Road Research Programme. At the time, all 19 Member countries expressed their intention to participate in the programme and finance it as a special budget. Today, the Road Transport Research Programme (RTRP) continues in this spirit as all 29 OECD countries are supporting Members of the programme.

Over the years, the Programme has become a true knowledge bank through a variety of mechanisms, including: an international network of experts and specialists in key transport issues; two operational international databases on documentation and traffic accidents; regular and cumulative enhancements of the research and development fields covered and policy challenges addressed; strategic communications with non member countries; and co-operation with other OECD Directorates and international organisations active in related areas.

The organisational framework is simple and clear:

- A policy committee - the "Steering Committee of the RTRP" comprised of senior research managers and administrators from Member countries;
- High-level seminars, conferences and workshops;
- Ad hoc scientific expert groups;
- Special joint research projects;
- Two functional, semi-independent data bases;
- A small, non-bureaucratic Secretariat team at OECD Headquarters.

This set-up assures focus and prioritisation that is responsive to the real needs of the transport and economic agencies in Member countries. A selective consensus on the investigations to pursue within the programme avoids overlap of efforts, nationally and internationally, and contributes to the economic use of scarce resources. Joint fact-finding and research exchange minimise the risk of formulating ineffective or irrelevant conclusions and recommendations in Member countries. The cooperative research projects and policy reviews are well-defined and carefully directed to assure results-oriented completion within a strict time schedule. This time schedule is generally 18 months or, for the larger and more complex research initiatives, three years.

The effectiveness of the Programme can be measured by the support and participation of Member countries. Since 1967, Member countries have exhibited their interest through continuing involvement in the RTR Steering Committee by their research leaders. Bi-annual Steering Committee meetings help shape and direct the Programme. In addition, and on an ongoing basis, representatives from 40 Member and associated institutions are involved in the development, update and promotion of the International Road Research Documentation and the International Road Traffic Accident Database. Also, over 30 years, more than 3,000 scientific, engineering, and economic experts have participated in and contributed to the investigative activities of the Programme. All of this points to the service orientation of the Programme and to the significant and meaningful role that the RTR Programme plays for Member countries.

I.2. WHAT IS AT STAKE

The Programme responds to the importance of roads and road transport to OECD economies. Roads are a major national investment. In the same vein, they exert a high cost to society, as measured by traffic accidents or environmental and energy costs. As an example, Table I.1 shows the total 1993 annual investment in roads for select Member countries. Figure I.1 indicates the estimated economic contribution of transport, storage and communication and highlights the magnitude of employment in the transport sector. In the EU for instance, road transport activities involve an estimated total of 6.5 million jobs.

Though the information presented in Table I.1 and Figure I.1 is useful, it does not tell the whole story. For example, most OECD Member countries have mature road systems that require regular maintenance investments to ensure they remain useable for the travelling public. Based on the report entitled *Road Maintenance and Rehabilitation: Funding and Allocation Strategies* (1), on average, OECD Member countries spend approximately 23 percent of their road budgets on maintenance activities.

Table I.1. **Annual total investment in roads (1993, $US billions)**

Japa n	Franc e	United Kingdom	Ca nada	United States
130	18.9	9.8	8.0	82.9

Source: U.S. FHWA and Japan MOC

Figure I.1.

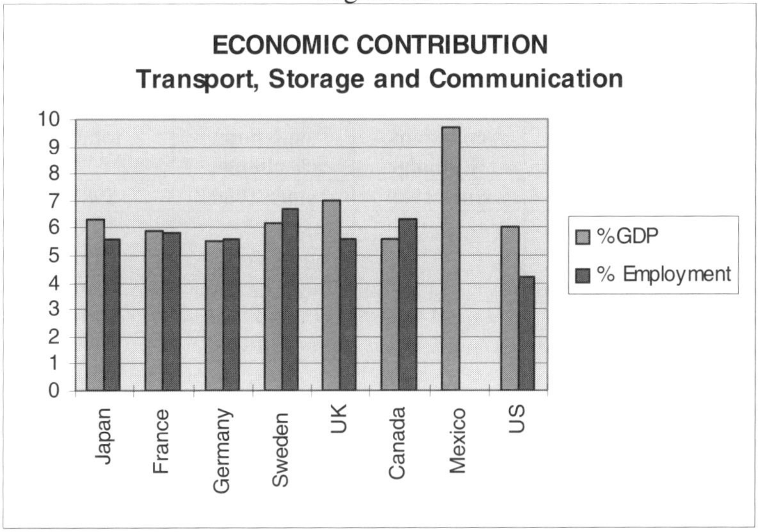

Source: U.S. FHWA

Unlike the infrastructure-related costs of road transport, other costs related to traffic accidents, the environment and the consumption of energy are not as plainly evident. For instance, the report entitled *Roadside Noise Abatement* (2), groups environmental noise costs as either the cost of prevention and remedy or social costs. In order to just correct existing roads in all OECD Member countries to meet acceptable noise standards would cost US$ 437 billion.

Another environmental factor associated with road transport is its impact on air quality. Table I.2 shows the impact of road transport on air quality as measured by its contribution to CO, CO2, and Nox emissions. The important points illustrated are: i) road transport is currently the overall single largest contributor to air pollution; and ii) that as a share of total air pollution, the contribution by road transport is generally increasing while the shares of all other sectors are decreasing.

Given that noise and air are only one part of the overall road environmental impact, it is easy to see how great the total environmental costs can be. The Programme is keeping countries engaged in this important area. For example, performance measures for the road sector were developed in 1996 and are to be tested in 1997/98 by a series of field tests (see Chapter IV). As well, a Scientific Expert Group is producing a report on *Integrated Safety/Environment Strategies* (3) that will be published in late 1997 (see Chapter VI).

Table I.2. **Contribution of the transport sector to selected air emissions**

	Total emissions % change	Transport emissions % change	Share of road transport in total emissions (%)	
	since 1980	since 1980	1980	mid 1990s
NOx emissions				
North America	1	-4	49	47
OECD Europe	-3	19	50	62
OECD	-1	4	49	51
G7	-2	1	49	50
CO2 emissions				
North America	10	21	22	24
OECD Europe	-5	48	14	22
OECD	6	31	18	22
G7	3	28	18	23
CO emissions				
North America	-14	-16	82	80
Europe (FRA,GER,ITA,UKD)	-9	-22	81	70
OECD	–	–	–	–
G7	-13	-17	82	78

Along with environment, one of the most significant cost factors to consider is road traffic accidents. It is estimated that the economic losses due to accident fatalities, injuries and damage amount to between 1 and 2 percent of GDP in OECD countries. Even more telling than this figure is to consider the cost to society of fatal accidents. A 1997 report by the European Transport Safety Council (4) suggests that the total socio-economic costs of transport accidents amount to over 166 billion ECU annually which is two times the 1997 European Union budget for all activities. In addition, as Table I.3 shows, road accidents account for about 97 per cent of the total costs. It should be noted that the figures in Table I.3 are estimates derived from using the "willingness-to-pay" approach. In the willingness-to-pay approach, people are asked how much money they would pay to reduce the risk of being involved an accident. The answers are averaged over a large population and result in an estimate of the social costs associated with adopting measures that will result in saving one statistical (not specific) life. Society-wide economic costs such as medical costs, net productive losses, and other costs are then added and result in the figures shown in the table. From these figures, it is clear that traffic safety is both a social priority and an economic challenge.

Table I.3. **Socio-Economic Costs of Fatal Accidents**

Transport Mode	Total socio-economic cost per fatality x million ECU	Estimated number of fatalities in 1995	Total socio-economic costs x billion ECU
Road	3.6	45,000	162.00
Rail	2.1	1,300	2.74
Air	2.7	186	0.50
Water	9.8	180	1.78

As a final measure, consider the use of energy. Figure I.2 (5) shows that for petroleum products consumed in OECD countries by all sectors, about 50 percent is consumed by road transport.

Figure I.2.

Total Petroleum Final Consumption

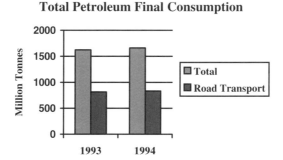

These data are presented to illustrate that the the costs associated with road transport in terms of the loss of human life, the impact on the environment, and the utilization of non-renewable resources are profound. In this regard, all countries, but especially the OECD Member countries, share a common interest in finding the means to reduce the high costs associated with road transport..

The RTR Programme is a fundamental partner of the transport agencies in Member countries in ensuring that road investments are made wisely, that economic contributions to transport are effective, and that road transport is as safe and environmentally sound as possible. It is in this context of major shared challenges that the OECD Road Transport Research Programme operates. As long as the OECD Member countries are faced with common problems on critical transport issues, it is in their benefit to work together to find solutions. The Programme RTR provides an effecient mechanism for this co-operation.

I.3. SHARING R&D AND BEST PRACTICES

The primary driving force for the Programme is the recognition by Member countries of the common challenges they share in the implementation of key transport and infrastructure policies and programmes. Whether considering the performance of road transport systems and road facilities, the maintenance of existing infrastructure, traffic safety, energy or environmental impacts, road and transport administrators throughout the OECD countries can identify their common ground. Thus, the Programme provides a productive and positive forum where these issues can be addressed co-operatively.

In spite of the central strategic value of the Programme, the Steering Committee has been compelled to note that research and development has not always received adequate attention within the budget or priority setting cycles in individual countries. Restructuring and down sizing of public research institutions has taken place in some countries and at the same time the private sector does not seem to have met their challenge. In a climate where needs outstrip resources, the participants in the OECD studies and co-operative research programs can obtain results that can either be applied immediately or adjusted conveniently in their countries to improve road transport.

Through participation in the OECD studies and seminars Member countries have an in-depth background and understanding of research performed in other countries that will serve as a benchmark for their own national and bi- or multilateral research programmes. Moreover, the Programme's two databases assure the systematic exchange of experience and research results so that all Members countries can benefit directly in terms of cost savings for research and improved systems.

Member countries also share a common commitment to exchange information with non-member countries. In the past three years, the Programme has supported Member countries in meeting their commitments through outreach efforts directed at co-operation with OECD non-member countries through technology transfer activities. The principal thrust of the activities undertaken is generally along the lines of providing technical or policy advice. Regional safety conferences in Asia, Latin America, and Africa are good examples of how the Programme works to bring state-of-the-practice information from Member to non-member countries to address a technical area that can have a profound impact on the targeted economies.

Meetings, studies, joint projects and conferences are meaningless until their outputs find their way into the rule making and working environments of Member countries. It is this philosophy that has guided the Programme to constantly have a process that is focused on results. In all of the approximately 150 scientific expert groups that have been convened by the Programme, the members are consistently encouraged by the OECD Secretariat to consider both the policy and technical implications of the work they are performing.

On the technical level, the final results of OECD/RTR reports identify best practices resulting from the implementation of research results in one or more of the Member countries. Likewise, they highlight areas where current research has or will make breakthroughs that can be of potential benefit to all Members. Finally, they indicate where further research by Member countries will have the greatest benefit in a given technical area. Through this process the Programme has helped magnify research and development results.

On the policy side, the OECD Steering Committee recognises the critical importance of examining the implications of multilateral initiatives. For this reason, all scientific experts groups are challenged to consider the policy implications of their work and to either report these implications or make recommendations that will assist Member countries set policies for their own situations. When projects address sensitive issues or the policy implications are serious, the RTRP Steering Committee is called upon to assist in addressing the policy questions that arise. Thus, on several occasions the Programme has been able to contribute to the policy-making efforts of Member and non-member countries alike and to regulatory work of the relevant international organisations.

I.4. STRATEGIC RESEARCH PLANNING AND EVALUATION

International co-operation in research has become much more of a strategic element. Most major and medium-sized OECD countries have a solid research infrastructure, up-to-date facilities and a well functioning science system. But, policy makers ask increasingly whether the existing system corresponds to the needs of the next ten to twenty, or even thirty years. Of course, at the same time, short-term policy research priorities cannot be discarded. The Road Transport Research Programme has therefore systematically considered the generic issues of research management and policy in a changing international context.

As a follow-up to the first Seminar held in the United Kingdom in 1990 on *Adapting Research Management to Future Needs* two similar Research Seminars were held during the 1993/96 period:

- Seminar on *Strategic Planning for Road Research Programmes*, Williamsburg (United States), October 1993;
- Seminar on *Development and Evaluation of Road Transport Research Programmes*; Annecy (France), October 1996.

Strategic planning for road research in the era of intermodal transport and international co-operation is vital. Anticipating future problems and timely development of new research measures can have significant economic benefits for the entire road community and its users. The main recommendations of the Williamsburg Seminar are summarised in "The Williamsburg Covenants" (Box I.1) that should be considered as guidelines for strategic planning of research programmes.

The Annecy Seminar was directly linked to the Williamsburg covenants emphasizing the need for evaluation. Indeed research evaluation is a key issue for research managers as it is the basis for the development of new research programmes and the implementation of potential results. The Seminar also stressed the role of public/private partnership and the benefits and requirements of international co-operation. Box I.2 summarises the main principles agreed upon by the participants at the Annecy Seminar.

Box I.1. **The Williamsburg Covenants**

1.   Research programmes should be planned with full consideration of "top-down" (policy) as well as "bottom-up" (technical) needs and initiatives.

2.   Research programme development is a dynamic process which must involve both the customers and the researchers.

3.   The expected benefits of the research programme should be identified in quantitative or qualitative terms (pre-evaluation).

4.   National goals for the research programmes should be established, and longer-term performance of the programmes should be measured against these targets.

5.   National research managers need discretionary funds to serve as seed money for future projects and to maintain the competency of the national research capability.

6.   Both national and international co-operative research should be encouraged in order to increase involvement and perspective and to leverage the limited funds available.

7.   Competition among research institutions invaluable, but all must be allowed to compete on equal terms.

8.   Applied research is important to meet immediate customer needs, but national programmes must include a longer-term perspective to identify and address future needs.

9.   Stagnant research programmes diminish a nation's research potential; dynamic research programmes nurture a national research capability and productivity.

10.   Implementation is critical for the success of research endeavours, and must be a continuing element of the research process. Technology has benefits only when it is put into practice. The linkage between research and technology transfer is essential for success.

Clearly, research is in constant evolution and new issues are always arising. The funding for research, its organization and the quality of the research itself are problems which most research institutes have to deal with. Because of new relationships with the private sector, new challenges need to be faced, including: Should the research institutions target public (governmental) needs or market driven priorities? How should public vs. private (user) priorities be reconciled? The mandate of the RTR Programme, and indeed that of the OECD Directorate for Science, Technology and Industry, calls for leadership in international discussions on these issues. The prime concern of the Programme, however, will be that the research activities undertaken are not separated from the real professional needs.

Box I.2. **The Ten Principles of Annecy**

Preamble

For a research programme to be successful, the creation of an organisational climate adapted to the R&D process is needed: clear mission; support and involvement at high level; stable funding; adequate facilities; high competence, creativity and diversity of researchers. Transparency and visibility are crucial criteria. Emphasis should be laid on creating new generic knowledge.

The R&D strategy in a targeted sector has to meet the vision assigned to this sector but also add a vision that underscores prospective scientific, technological and societal trends.

Public agencies should conduct anticipatory research, to aid in international bench marking and support visionary transport policies. The open exchange of international experience is essential.

Evaluation

1. Evaluation is a critical issue for all types of research programmes and projects. The evaluation may be ex-ante, concomitant (mid-term) and ex-post. It may focus on the development and management of successive research phases and involve impact studies. Evaluation may cover the whole process from the development of the programme up to the implementation of results.

2. The evaluation must have clear objectives. The evaluation process must take into account the degree of risk associated with the research undertaken as well as the type of research prevalent in the innovation process: incremental; breakthrough; normative or regulatory; technology transfer, training and education.

3. The evaluation must be conducted rigorously and designed, implemented and used not only for ex-post judgment, but also to formulate and manage future policies and programmes.

4. The evaluation during the realisation of a programme (or project) permits the redirection or redefinition of the work to meet programme needs and specific customer requirements and values.

5. In the case of short term, ex-post evaluation, efforts to demonstrate the overall productivity or effectiveness should be made periodically.

6. Creative and rigorous methods are needed to evaluate the potential value of anticipatory research, which should be judged both in the short and the long term.

7. The evaluation must be credible undertaken by reputable evaluators who respect the confidentiality clauses, and provide intellectual fairness and integrity. In various cases, international peer reviews can be useful. An appeal procedure concerning the results of the evaluation must be set up.

8. Programme budgets must include sufficient funding to support evaluation, and be adapted to the size and importance of the research undertaken.

9. The evaluation should champion more creativity in research, including creation of knowledge networks and infrastructure such as centres of excellence.

10. Evaluation guidelines setting forth principles, targeted standards and quality assurance goals would be useful.

I.5. SHIFT FROM TRADITIONAL RESEARCH TO NEW FRONTIERS

Admittedly, it is difficult to assess the effectiveness of international co-operative research. The following examples are presented to illustrate how the service and results orientation of the Programme has stimulated progress, new thinking, and changes in professional practice.

At the instigation of the Australian Delegation, the Dynamic Interaction Vehicle-Infrastructure Experiment (DIVINE) was launched by the RTR Steering Committee. Through this action the Steering Committee acknowledged the relevance of previous OECD efforts on heavy vehicles, pavements, and bridge performance. Likewise, the Member countries indicated their belief that a major research effort such as this required the combination of the unique talents, capabilities, and facilities found in individual countries. Beginning with financial contributions from 17 countries and the European Union amounting to $US 1,500,000, the Members committed staff time, resources, and facilities.

The combined effort produced significant technical advances in how Member countries respond to the impacts of heavy vehicles on infrastructure. Because this is an extremely sensitive area for many countries, the policy implications are greater than with many other research activities. Therefore, the OECD Steering Committee developed a policy note that is apart from the technical reports and recommendations. They will be reviewed in 1997 at high-level workshops in Canada, Europe (Netherlands) and Australia. The OECD considers the DIVINE project to be a showcase in bringing diverse interests into a harmonious environment to advance technical and policy thinking on subjects that can potentially have a significant impact on Member country economies and will shape new reseach directions.

Beginning in 1976 with a report on bridge inspection, the OECD has been impacting bridge programmes in several countries. For example, as a result of the 1976 study, the United Kingdom, Sweden, Finland and Indonesia all adapted their standards and procedures. The Programme then pursued other bridge areas by dealing with *bridge maintenance, bridge rehabilitation and strengthening, durability of concrete structures,* and more recently producing a report on *Repairing Bridge Substructures* (6). The 1992 report entitled *Bridge Management* (7) was the key reference document for the International Seminar on Bridge Engineering and Management in Asian Countries held in September 1996 in Jakarta, Indonesia. In summary, the OECD bridge reviews have had a marked impact in the maintenance field and have been appreciated for their practical and guiding value by the technical agencies in OECD Member countries.

In transport telematics, the OECD was at the centre of an in-depth, tri-lateral (U.S./Canada, Europe, and Japan/Australasia) co-operative effort that began right at the inception of the Programme in 1967/68. Scientific Expert Groups reported on *Electronic Aids for Freeway Operation* (8), *Area Traffic Control Systems* (9), and *Research on Traffic Corridor Control* (10). The information developed in these efforts was shared and enhanced at international seminars in Toulouse, Hamburg, and Aix-la-Chapelle. Some of the primary impacts resulting from the OECD work included the development of criteria and definitions for incident management, guidelines for the design of variable message signs, and a protocol for performing joint international research. By the end of the 80s, the OECD completed three reports on *Dynamic Traffic Management in Urban and Suburban Road Systems* (11), *Route Guidance and In-car Communication Systems* (12), and *Intelligent Vehicle Highway Systems* (13). The work of the OECD continued to contribute to global (trilateral) co-operation as independent national and international groups have emerged to expand the state-of-the-art in this area.

As mentioned earlier, the Programme also seeks to transfer technology to non-Member countries through its outreach activities. For example, the Programme's ability to harness diverse resources from Member countries and apply them to specific regions has had a meaningful effect in the Central and Eastern European Countries (CEECs). The OECD has now hosted one introductory seminar (14), 14 workshops (15) and a concluding conference (16) for the CEECs. Through these intitiatives, the Member countries have provided meaningful tehnical and policy support to these countries in transition. The full impact of these efforts are yet to be known, but already the infomation provided has motivated some countries to adopt new policies and new practices that are generating benefits in terms of safer, more efficient, and environmentally friendlier transportation.

I.6. MEETING THE CHALLENGES OF THE FUTURE

In the same manner that the OECD and the Steering Committee must look back to gauge results, so must they continually look ahead and consider what factors may impact road programmes in the Member countries and elsewhere. Through these considerations, the RTR Programme can develop and design activities that will address both the immediate needs of the OECD Member countries as well as contribute to future challenges that will likely confront these same countries as transport demands change. Three of the primary trends that will affect the conduct and output of the RTR Programme over the next triennial period are the globalisation of trade, the application of advanced technologies in transport, and the increasing requirement for intermodal connections between road systems and other forms of transport. All of these subjects are covered in later chapters.

This 30th Anniversary report is designed to provide a good understanding of the activities encompassed by the OECD Road Transport Research Programme. It is organised by technical groupings as follows: Intermodality and Logistics; Advanced Traffic Technologies and Practices; Infrastructure; Road Safety; Environment/Transport Interaction; Technology Transfer Activities and Two Databases. Many of the subjects that were briefly touched upon in this chapter are explained in more detail in the ensuing chapters. After studying this report, the reader should be left with a sound understanding of the continuing contributions of the RTR Programme and the prospects it holds for bringing focused international attention to the pressing transport needs facing the OECD Member countries and all countries world-wide.

I.7. REFERENCES

1. OECD ROAD TRANSPORT RESEARCH PROGRAMME (1994). *Road Maintenance and Rehabilitation: Funding and Allocation Strategies*, OECD, Paris.

2. OECD ROAD TRANSPORT RESEARCH PROGRAMME (1995). *Roadside Noise Abatement*, OECD, Paris.

3. OECD ROAD TRANSPORT RESEARCH PROGRAMME (1997). *Integrated Safety/Environment Strategies*, OECD, Paris.

4. EUROPEAN TRANSPORT SAFETY COUNCIL (1997). *Transport Accident Costs and the Value of Safety*, ETSC, Brussels.

5. INTERNATIONAL ENERGY AGENCY (1996). *Energy Balances of OECD Countries: 1993 - 1994*, OECD, Paris.

6. OECD ROAD TRANSPORT RESEARCH PROGRAMME (1995). *Repairing Bridge Substructures*, OECD, Paris.

7. OECD ROAD TRANSPORT RESEARCH PROGRAMME (1992). *Bridge Management*, OECD, Paris.

8. OECD ROAD TRANSPORT RESEARCH PROGRAMME (1971). *Electronic Aids for Freeway Operation*, OECD, Paris.

9. OECD ROAD TRANSPORT RESEARCH PROGRAMME (1972). *Area Traffic Control Systems*, OECD, Paris.

10. OECD ROAD TRANSPORT RESEARCH PROGRAMME (1975). *Research on Traffic Corridor Control*, OECD, Paris.

11. OECD ROAD TRANSPORT RESEARCH PROGRAMME (1987). *Dynamic Traffic Management in Urban and Suburban Road Systems*, OECD, Paris.

12. OECD ROAD TRANSPORT RESEARCH PROGRAMME (1988). *Route Guidance and In-car Communication Systems*, OECD, Paris.

13. OECD ROAD TRANSPORT RESEARCH PROGRAMME (1992). *Intelligent Vehicle Highway Systems*, OECD, Paris.

14. OECD ROAD TRANSPORT RESEARCH PROGRAMME (1991). *Seminar on Road Technology Transfer and Diffusion for Central and East European Countries*. Budapest. 12th-14th October 1992.

15. OECD ROAD TRANSPORT RESEARCH PROGRAMME (1995). *Road Infrastructure Rehabilitation and Safety Strategies in Central and Eastern Europe*, OECD, Paris.

16. OECD ROAD TRANSPORT RESEARCH PROGRAMME. *Concluding Conference of the CEEC and NIS Series of Workshops*. Ljubljana, Slovenia. 19-20 October 1995.

CHAPTER II: INTERMODALITY AND LOGISTICS

II.1. RESPONDING TO GLOBALISATION TRENDS

Globalisation and the rapid improvement of information and communication technologies are profoundly affecting the infrastructure and transport policies of OECD and other countries worldwide. Or conversely – as General Secretary Donald Johnson put it at the December 1996 WTO Ministerial Conference in Singapore -- *Transport and communications technologies have been leading the way.... in our age of globalisation.* Transport is an essential factor of economic activity, industry and trade and therefore the Programme's mission is clearly in step with OECD's objectives to achieve the highest economic growth, contribute to the development of the world economy and to expand world trade.

The globalisation of trade is an issue that is affecting all countries. The increasing number of countries adopting market economies has brought about a sea change in how countries view the potential for international commerce. The diversification, specialisation and integration of markets and the potential and impact of emerging or changing patterns of globalisation in the OECD regions have added a new dimension to freight transport and affected the structure and operation of the transport industry as a whole.

Additionally, the creation of regional trading pacts and partnerships has made international trade a critical factor in the economies of most countries. Free trade relies on the ability to move people or goods from one place to another. This requires adequate and robust transport systems nationally, regionally and internationally. It calls for the harmonisation of technical standards between partner countries. But it also means that, gradually, the prioritisation and co-ordination of transport plans and programmes must take into account the requirements of a partner country. Without such assurances, the safe, fast, and economical transport of people and goods is far less likely to be available to support increased trade and prosperity.

With globalisation and the increasing need for competitiveness, the ability of countries to reduce transaction costs through the provision of adequate and efficient *intermodal transport systems* is more critical than ever. The OECD understands these relationships and the importance intermodal connectivity plays in the provision of road systems. Clearly, roads are not alone in the transport of people and goods. Though the land border crossing facilities are the major gateways for people and goods to flow between trading partners who share a land border, roads also serve trade that arrives or departs at airports, seaports or rail freight terminals.

Spurred by the global market place, there have been profound changes in the production system and physical distribution. *Advanced logistics* is now fully recognised as a crucial motor in the competitiveness of industry and trade companies where transport is a major element of the production-distribution chain. The value added by transport depends on the extent to which transport services are integrated with other logistics functions. All forms of transport contribute to added value, and the value added by transport can be very high if the operations take place in a well designed and market responsive intermodal system. Information technology and communications networks are crucially important in enhancing transport systems and improving the performance of freight transport services.

II.2. MULTIMODALITY/INTERMODALITY: KEY CONCEPTS IN MODERN TRANSPORT POLICIES

Governments are deeply involved with planning, investment, construction, and/or operational decisions that influence the provision of capacity and the pricing and time dependent availabilities of transport facilities and services. For almost all modes today, the levels of congestion and delays in transport and at terminals are an important problem. A bottleneck or a missing link results in less efficiency for the whole system. One small link investment can therefore have a dramatic effect on transport efficiency, while another only creates overcapacity. What counts is the performance of the whole network of infrastructure not the performance of individual links.

An intermodal assessment and systems approach can therefore provide long term economies and efficiencies. Intermodalism is now a prime policy objective in many OECD countries including the United States and the European Union.

However, it should be pointed out that the need for a more comprehensive "multi/inter"-modal[1] approach to transport policy and infrastructure development is not considered equally urgent in all OECD countries.

In many European countries, the need for an intermodal transport policy is strongly advocated because of environmental concerns, reasons of overall efficiency and the benefits of co-ordination of modes to cope with growing transport flows. In the United States, intermodal transport options are developed not only because the administration is in favour of them, but also because they are considered to be cost-efficient in the North America context. In Asia and Australia, countries are still building their unimodal transport systems, and therefore do not yet feel the need for an overall multi/intermodal transport policy.

II.2.1. Exploring priority setting

The OECD Seminar on *Prioritisation of multimodal transport infrastructure*, placed under the leadership of The Netherlands, took place in May 1996, with the co-operation of the EU (1). Transport infrastructure considered included roads, railways, inland waterways, seaports and major airports, with the nodes and links interconnecting them. The focus was on methods and procedures for appraising multimodal infrastructure alternatives and on ways of incorporating aspects of sustainable development such as environmental quality and external costs, in general.

The major achievement of this OECD project was to set the grounds for a framework for evaluating and prioritising multimodal infrastructure programmes and alternatives. The Seminar provided the elements of a strategic planning

---

[1] A number of definitions in relation to freight transport can be proposed:

- **Intermodal:** the movements of goods which use successively several modes of transport without handling of the goods themselves in changing modes;

- **Multimodal:** carriage of goods by at least two different modes of transport in succession;

- **Combined transport:** initial/terminal transport by road - major transport by rail, waterway or sea.

In this section of the report, "multimodal" and "intermodal" are used as synonym to avoid semantic discussions.

approach with due regard to land use, environmental and industrial – logistics – goals. In addition to the technological and strategic dimensions, it was clear that the financing issues dominate the development of intermodal networks and facilities. Box II.1 contains the recommendations of the OECD Seminar highlighting major elements for promoting multimodal developments.

Discussions showed that the dialogue among transport agencies (including roads, railways, coastal shipping, waterways and air) should be enhanced in order to ensure cost-effective co-ordination of transport modes. There is also a growing belief that the traditional means for funding massive infrastructure programmes are no longer adequate in meeting global trade developments, spurring economic growth and creating jobs. In this regard, there is a need for institutional innovation in increasing collaboration between the private and public sectors. As shown later in this chapter, the logistics sector can serve as an efficient paradigm for public/private infrastructure initiatives.

II.2.2. Developing intermodal systems and financing

The OECD Conference on *Intermodal Transport Networks and Logistics* to be held in June 1997 in Mexico City will cover most of the development, financing and operational issues of intermodal systems based on practical experiences gained in Member countries. A fact finding mission has therefore been entrusted to The Netherlands Ministry of Transport, Public Works and Water Management, whose Strategic Policy Group is currently conducting an international survey on behalf of the OECD. The review centres on policies, decision-making processes, technologies, logistics, market organisation, legislation, financing and future national agendas. This should provide an overview of success stories, failures, bottlenecks, constraints and expectations of intermodal networks.

It is the increasing importance of logistics, and the emerging development of intermodal systems – together with new financing mechanisms – that prompted the Mexican Ministry for Transport and Communications to prepare for this Conference in co-operation with RTRP and involving NAFTA, APEC and the Latin American Transport Ministers. It is recognised that the vested interests of the public authorities to meet economic and policy goals can be usefully linked to the requirements of the private sector to meet business targets. Clearly, the strategic parameters need to be assessed especially in regard to the competitiveness goals of firms and financial public/private outlays.

Box II.1. **Recommendations for progress on multimodality (4)**

1.   Process and method of decision-making

- Transparency in the decision-making process is needed to gain support for multimodal programmes and their implementation;

- Independent liaison units to conduct the dialogue with the public are advocated;

- Methods and procedures for multimodal appraisal and priority setting should combine the more traditional cost-benefit method and the "partnership method" being used in some metropolitan areas. In the "partnership method", planning, project development, implementation and system management are handled collectively by various government agencies and service providers.

2.   Administration

A multimodal approach of planning should be reflected in the organisational structure of the government agencies involved, aiming, if need be at:

- a change of culture of the agencies involved;

- the decentralisation of responsibilities;

- the enhancement of co-operation between government agencies.

3.   Professional networking

The multimodal approach requires an international network of professionals active in multimodal transport and logistics, including policy and decision makers from central and local government agencies, transport and logistics providers, financial and infrastructure operators and industrial users of transport and logistics.

4.   Technological improvements

There is a range of proven technological improvements available, with many opportunities to upgrade the logistics chain. Many of theses technologies have already been applied and may penetrate further into practice during the next few years. Information Technologies (IT), especially, are considered to become pervasive.

5.   Research need

The multimodal approach requires internationally accepted multiple criteria to support the decision-making process.

Box II.2 highlights the Mexico Agenda. The conceptual presentations at the Conference are complemented by a series of intermodal case studies from Singapore, Rotterdam, Roissy, Transportación Marítima Mexico, the US Landbridge, Japanese industry, the Unification Projects of Germany, etc. Experts from governments, academia, industry and the banking sectors have agreed to participate.

Box II.2. Mexico Agenda

Objectives

- To present progress and experience on intermodal transport highlighted through OECD reviews and actual cases worldwide;

- To contribute to the development of policies and strategies of intermodal transport systems with a view to roles and responsibilities of public and private entities at the national and regional levels;

- To promote new public/private partnerships to develop transport infrastructure and financing systems;

- To develop an international network for professionals active in intermodal transport and logistics, including policy and decision makers from central and local government, transport and logistics providers, financial and infrastructure operators and industrial users of transport and logistics.

Emphasis areas

- Setting the scene of intermodality and logistics worldwide;

- Development of multimodal transport: phases, problems and opportunities;

- Multimodal strategies for transport development;

- Multimodal facilities and operation: 10 selected case studies;

- Innovative financing of transport infrastructure from a modal and multimodal approach;

- Integrated advanced logistics and multimodal transport.

Generally, it is felt that closer contact and dialogue with the banking sector should be pursued more rigorously in the future. Governments in industrialised economies, economies in transition and in developing countries are finding it increasingly difficult to construct and maintain their own transport infrastructure system. It is recognised that governments should therefore permit maximum flexibility in the use of alternative funding sources. In order to evaluate the extent of the problem, a dedicated session of the Conference will discuss the potential of innovative financing methods from a modal and multimodal approach.

It is important to debate and compare experiences in different countries as well as the current trends in financing mechanisms in the world. Special attention will be given to private sector involvement, and especially the vision of commercial banks on infrastructure project financing. The support of international financing institutions to transport infrastructure projects under mixed public and private financing schemes will also be evaluated. National case studies and review of practices will help provide answers to key financing questions such as: Are guarantees needed to ensure that viable projects are financed with private funds? how should risks be distributed among partners? what actions should a government take to facilitate different financing schemes for infrastructure projects?

II.2.3. Global logistics and intermodal transport

Logistics channels that are more global also require multi/intermodal transport chains. Intermodalism is at the core of most advanced logistics strategies used by the major transport companies in the world. Intermodal management responds to the challenges faced by logistics service providers. The co-ordination of production and distribution in an integrated process is becoming of strategic importance to many companies. The logistics service providers must be able, first and foremost, to provide a timely, cost effective, high quality service to their customers and therefore operate in an intermodal environment. Clearly, the terminals for trans-shipment between modes are becoming increasingly important.

To illustrate trends, take the case of UPS which is the world largest package delivery company. In the beginning, UPS started out with foot messangers, then they were put on bicycles, and later at the wheel of a fleet of trucks. Those trucks got bigger and were piggy backed onto railroad cars and, in order to cover vast distances, UPS launched its own airline. This shows that the system of transport must be transformed and adjusted to meet the requirements of the worldwide market place in order to maintain competitiveness of companies. Yesterday's customer told carriers how they wanted goods moved, today's customers tell carriers when they want goods to get there and do not care about the mode selected to move it as long as they are satisfied.

Also, new infrastructure-related initiatives are needed. The fast rail infrastructure now being developed for passenger transport should be partly usable for freight as well. High-speed combined transport trains are now being tested up to speeds of 240 kilometres per hour in Germany, and are commercially operated at 160 kilometres per hour in France. High-speed freight trains also increase the capacity of the rail system, since these trains do not have to be shunted aside to make way for faster passenger trains. Note that the current average speed of rail transport from, for example, Germany to Italy is about 18 kilometres per hour (2).

In countries like The Netherlands, Austria, Germany, Switzerland, and France with heavy truck through-trips, there is a strong declared intention to improve combined transport facilities primarily for environmental and economic reasons. However, in most cases and with the exception of Switzerland, there is a shortage of realistic political support programmes to implement a daring strategy. The key to increase the existing combined transport facilities lies largely with logistics. It is, however, necessary for railway authorities to give this subject a greater priority than present tentative development studies currently imply (3).

II.2.4. *Perspective*

The shift to intermodalism requires a revolution and a change in mentalities. Intermodal aspects must be brought in at all phases of transport policy and planning, including R&D. Advanced logistics have shown to be effective in promoting greater co-operation between transport carriers and competing modes.

It is envisaged to follow up the Mexico Conference by screening existing strategies and the experiences gained. The aim of a 1998 OECD study is to develop a conceptual framework for a forward looking approach which promotes intermodal transport and to specify the design principles for intermodal infrastructure, facilities and networks.

II.3. CHALLENGE AND OPPORTUNITIES OF ADVANCED LOGISTICS

As mentioned previously, the continuing trend toward globalisation of economies and markets is leading to profound changes in the production and physical distribution systems. The concept of *"advanced logistics"*[2] covers management of production and freight business planning and relates to the transport components and associated information flows needed to respond to changes in the demand for transport services. In advanced logistics systems, information technology plays a key role. In some ways, information technology is as important, or more important, than the product itself: in the case of just-in-time shipments (6) to a production line, it is almost more important to know that the goods will not arrive than it is to have the goods; at least, if there is sufficient warning of a problem, corrective actions can be taken.

Technological innovations create possibilities for significant changes in logistics. Many of these innovations occur in different countries or different regions thus calling for a joint assessment on a worldwide basis. In order to monitor these changes and their impact on future transport policies, attention was given in the 1996 report on *Integrated Advanced Logistics for Freight Transport* (4) to the effects of innovation on the competitiveness of companies as well as the role of governments in increasing opportunities for intermodal co-ordination to provide integrated logistics services.

Comparing the results of the 1992 OECD study on *Advanced logistics and road freight transport* (2) with those of the 1996 review (4), it is easy to see that the trend of introducing advanced logistics in industry and transport management shows no sign of slowing down. Firms are under increasing pressure to adapt to the latest innovations. However, the public sector plays an

---

[2] The following definition is proposed:

"Advanced logistics is the concept of synchronising the activities of multiple organisations in the logistics chain and feeding back necessary information to organisations in production and/or physical distribution sectors on a real time basis, by fully utilising information technology and digital communication networks" (1).

The commercial applications of *"advanced logistics"* have come to be called "supply chain integration". Some experts use the term "demand chain integration" to signify that the entire process is driven significantly by the demands (in the sense of consumption) of the actual consumers or users of the goods.

important role in the development and application of many of the technological advances, both as a direct user as well as a regulator or investor.

It is very important for governmental bodies to facilitate advanced logistics even if in some countries questions arise regarding the appropriate role and involvement of government in these systems. The 1992 OECD report (2) highlights the need for governments to actively participate in the spread of advanced logistics in ensuring open access and availability of information technology (IT) innovations to all potential users, including small and medium enterprises, while at the same time minimising inefficiencies that may be due to outdated government regulations. Co-operation and joint private/public activities and investments may become more common place in OECD Member countries in the future in order to serve the requirements of both sectors simultaneously.

II.3.1.   Driving role of IT logistics

The key dynamics of the development of modern logistics lies primarily in IT as an integrating tool and support of some basic and multi-organisation processes such as the interchange of information among trading partners in the supply chain or the management of multiple transport modes requiring new ways of co-ordination. IT and advanced logistics operations are inseparable.

Keeping in mind the need for higher quality source data to ensure efficiency and a high operational standard, technologies such as Automatic Equipment Identification (AEI), Global Positioning Systems (GPS) or selected components of Intelligent Transport Systems (ITS) are or will be useful. To help support inter-entreprise co-ordination through structured data such as Electronic Data Interchange (EDI) and to achieve a high degree of integration through IT, it is important to clarify the structure and relationships between freight transport and information networks.

Such analyses can be carried out through different approaches and the OECD study report (4) identifies an evaluation process from the macroscopic view of logistics. This includes not only the viewpoint of the transport industry but also the viewpoint of public interests taking into account environment, health and safety requirements. The method is based on the systems architecture – which describes the logistics systems – and the structure of information processing attached to the logistics operations.

Figure II.1 shows the hierarchical structure of information processing. The level of hierarchy in the information processing corresponds to the hierarchical structure of an organisation which can be a whole company, one division or an aggregation of various organisations. The hierarchical structure or the vertical structure of information processing shows how the data or the information are aggregated or accumulated to provide the necessary data for managerial decisions.

Figure II.1. **Hierarchical structure of information infrastructure for logistics systems and information processing in a managerial process**

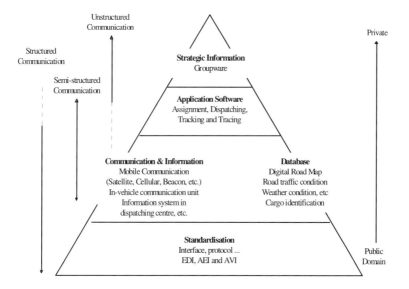

II.3.2. *Building systems architecture*

The overall architecture of a logistics system is composed of three architectures namely: logistics, institutional/organisational and information/ telecommunications:

- Logistics architecture is a complex structure in terms of flow of freight, operations, intermodal facilities, flow of the transaction of messages and data required for operating the system.

- Institutional/organisational architecture is a framework to explain or illustrate how actors and stake holders act in the process of improving the information infrastructure and how these actions influence the design of information systems.

- Information/telecommunications architecture is strongly dependent on the rapid development of IT and telecommunications techniques over recent years leading to the centralisation of resources and operators.

Logistics architecture

The complexity of logistics systems is due to the various existing forms of organisations which actually operate freight transport. The Role Player Model presented in Figure II.2 expressing the more practical aspects of logistics systems – flow of goods and flow of information – meets the requirements to be an essential tool for constructing logistics architecture.

Figure II.2. **Role player model of logistics operations**

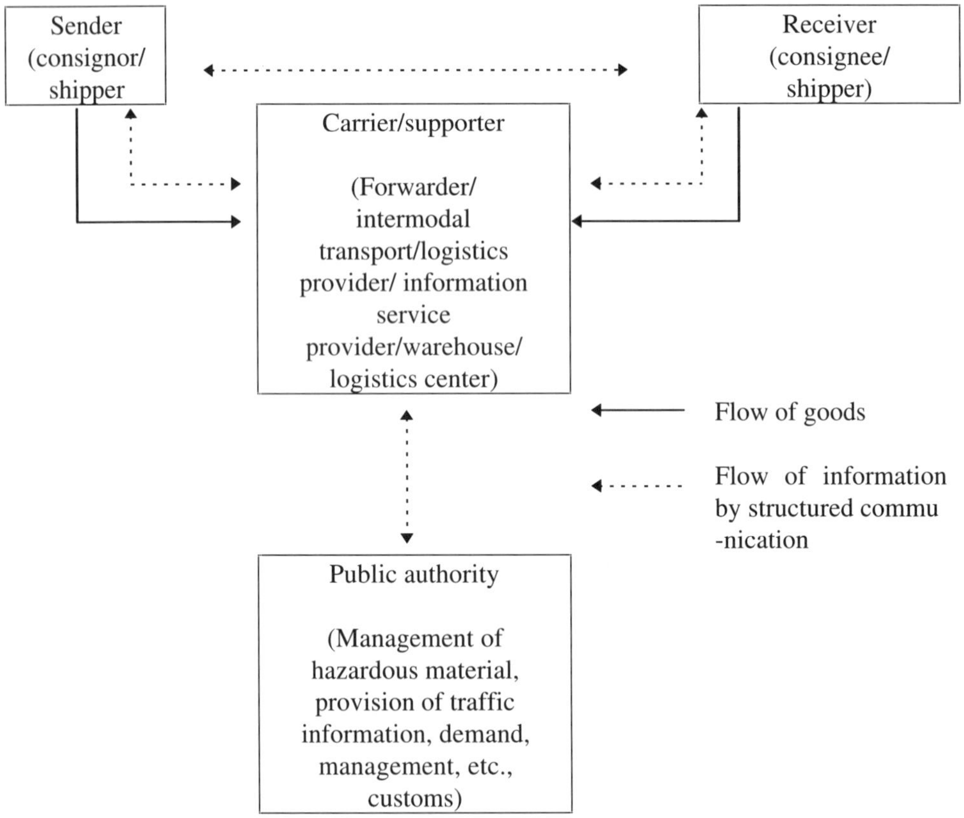

To illustrate the role player model, examples of Cash transport – Optimal dispatching system – Parcel delivery – JIT production – Public authorities and HAZMAT transport are presented in the 1996 report. They show that such models have different requirements for information technology due to the differences in the characteristics of freight.

Information and telecommunications architecture

Because of the rapid development of increasingly powerful information and telecommunications technologies, the integration in the logistics system of an efficient and reliable information and telecommunications architecture is not straight forward. The highly hierarchical information systems must be made compatible with the more horizontal and natural architecture of local networks. Gateways or bridges have to be foreseen to provide network interfaces (see Table II.1).

Computer applications rely increasingly on de facto standards and in particular on relational databases. With the advent of micro-computing, the principle of common use of central units has been questioned especially due to the widespread use of networking with which it is possible to access several processing facilities from a single work station. For example, INTERCONTAINER has many modern and efficient facilities for handling a full range of combined transport operations by using the various databases it manages as a client/server model.

Table II.1. **Examples of network interfaces**

Interfaces	**Number of networks**	**Comments**
bridge	2	passes data from one network to another
b-router	2	intermediate between bridge and router
router	2	filters messages and finds the correct route
hub	N	same for several networks
VAN server	N	processes part of the contents and stores the messages (letterboxes)

Box II.3. **Examples of activities related to institutional aspects**
(Co-operation between private and public sector)

Third party group

Various financial and legal regulations concerning the establishment, the organisation and the management of third party groups exist, but they are quite different from country to country and case to case. The Japanese VICS (Vehicle Information and Communication System) is an example. It is an organisation for the provision of traffic information in Japan. Three ministries in Japan -- Ministry of Post and Telecommunications, National Police Agency and Ministry of Construction -- are involved.

The communication links of VICS are based on FM multiplex, roadside beacons (microwave and infrared) and a cellular data transmission system. The traffic information accumulated by a traffic surveillance system is processed and transmitted to vehicles through VICS links. Drivers profit by having dynamic traffic information while driving the vehicles equipped with a VICS communication link. On the other hand, the road authorities benefit insofar as the road network is efficiently used thanks to the drivers who adjust to the information provided through the VICS communication link.

Development of new technologies for logistics

New technologies related to mobile communication and traffic management are under development in ITS programmes. Various forms of co-ordination between public and private sectors exist in the US, Japan and Australia. ITS related projects cover a wide spectrum of advanced transport systems and enhanced logistics are only one part of the overall plan.

The emphasis has recently been focused on intermodal aspects of transport systems. The advancement of logistics will become a very important part of research. Most of the outputs of ITS related R&D projects will have substantial impacts on future logistics.

Promotion of co-operative operation of physical distribution

The Japanese law "Physical distribution efficiency promotion act for small and medium enterprises" in October 1992 is a case in point. This act consists of measures to promote and support the introduction of co-ordinated operation of physical distribution. The operation is conducted by establishing an organisation -- called co-operative association -- which involves various industries, manufacturers, retailers, warehouses and forwarders.

The support is through subsidies, financial credits and loans, special tax reductions, investment, etc. The most interesting measure in this act is the special loan system to promote the construction of a co-operative physical distribution centre. The loan is applicable to funding the purchase of land, facilities and equipment for the construction. In such a case, the association that has a plan to construct a co-operative physical distribution centre can receive loans of up to 80% of the necessary funds without interest.

Another important aspect is the variety and complexity of the operation and development of communication facilities which require substantial administration in order to maintain a satisfactory level of availability, accessibility and reliability. Network administration has become primordial and plays a critical role. As a result, more and more interconnection platforms are being created to satisfy the interlinking requirements of companies' networks. TRAXON is a typical example in which three airline companies participate to exchange electronic information with forwarders.

Institutional architecture and system design

In most countries, freight transport and road infrastructure are dealt with by different ministries or divisions in a ministry. Innovations in information and communication technologies do not fit the traditional framework of jurisdictions but at the same time, the globalisation of the economy has led to the "borderless economy" for which the traditional process of government is not suited. Typical situations where institutional issues need to be clarified are, inter alia, cross border operations and management – openness, compatibility, interoperability and flexibility of systems to operate with various software and hardware – public/private co-ordination (Box II.3) and project implementation – city logistics.

To develop the most appropriate institutional architecture, the specific logistics problems and information processes need to be identified as well as the prevailing organisational structures. It is important to clarify the interrelationship with the other architectures. There is also a need for future research to compare the institutional frameworks among countries in order to extract those problems that are linked to the rapid development of globalisation and that are of crucial importance when establishing international frameworks and standards to enhance advanced logistics.

II.3.3. *Impacts of logistics on infrastructure policies*

Need for integrated investment programmes

Efficient logistics and transport operations require harmonious flows of goods and traffic throughout both the transport and logistics chains. While in the past, infrastructure investment programmes were developed from a single mode approach, global logistics operations today favour the integration of modes and impose new priorities in the decision-making process.

The logistics shifts towards global sourcing and centralised inventories demand larger geographical coverage and smooth international transport without any counterproductive delays in the network and at border crossings. Clearly, a bottleneck or a missing link results in less efficiency for the whole network. In fact, most "missing links" in Europe have been identified in border areas, e.g. the Alpine crossings, the UK Channel Tunnel, the Great Belt, Öresund and links with former Eastern Europe.

In Europe, the primary infrastructure problem is that investment in new infrastructure has been much less than traffic growth over the past ten years and, given current predictions of traffic growth, capacity needs to be doubled within 20 years. The Round Table of European Industrialists has concluded that the decision making process at all levels – local, national and international – is increasingly inadequate for creating, expanding and operating very large and complex transport infrastructure projects now required. There is a need for increasing collaboration between private and public sectors and institutional innovation in this field. The logistics sector can serve as an efficient paradigm for public/private infrastructure initiatives.

In the United States, the future freight movement policies are likely to be characterised increasingly by intermodal features governed by logistics needs. Major road corridor studies along the East, Midwest and West include provisions for improved truck flows and for better access to ports of shipping (rail, water and air terminals). New policies presented under the Intermodal Surface Transportation Efficiency Act of 1991 mandate that cities and States incorporate freight movement activities into their congestion management and overall transport programmes. Freight movement measures should be incorporated into congestion management activities as a matter of national and local policies.

In Japan, global logistics and freight transport operations have underscored the need to attenuate critical congestion problems in urban networks for the delivery of goods at the intended time, raising the issue as to whether a more efficient usage of transport capacity could be achieved. The creation of an underground network in Tokyo for the transport of freight containers is presently under study. The Japanese example clearly shows that traffic congestion is worsening in metropolitan areas calling for the implementation and reorganisation of urban deliveries requiring strong incentives by governmental bodies to help both shippers and carriers.

City logistics - Improving urban freight transport

Logistics services changed and progressed rapidly after the 80's from widespread Just-In-Time transport to consumer related transport services exerting pressure on traffic conditions, especially in urban areas. The limitations and constraints associated with urban freight distribution and/or logistics activities must be overcome, and in no case logistics activities should be a bottleneck for economic developments. Therefore, "city logistics" is very important, and the following proposed measures could be taken to improve the urban traffic environment:

- Change the form of urban delivery from independent transport using vehicles owned by each shipper to consolidated transport using public haulage. Decrease the number of delivery vehicles through joint delivery by several trucking companies having the same consignees or delivery areas.

- Create an Advanced Dispatching System in order to increase the loading rate of urban delivery vehicles and decrease inefficient transport. (Such systems require the application of advanced information systems for dispatching and vehicle scheduling, such as computer aided dispatching systems using mobile data communication systems, positioning systems and digital road maps).

- Rationalise transport and delivery planning systems. Effective utilisation of urban parking facilities in order to decrease on-street parking for the loading and unloading of cargo in urban areas.

Logistics terminals and platforms

An organisational structure such as "platforms" or "logistics terminals" can be considered a relatively recent development in the road transport sector although these existed for other modes for a long time. In order to respond to economic and productivity constraints associated with the need to improve traffic conditions at some locations of the network, or more especially in urban areas, the spatial planning of platforms must be carefully evaluated. Their location is directly dependant upon infrastructure capacity including information and telecommunications potential and the economic activity of the region.

For these reasons, siting and development of logistics terminals and platforms cannot be discussed separately from infrastructure planning. Long term public infrastructure planning is, hence, closely interrelated with private sector initiatives which, however, are more business and short term oriented. The challenge for future infrastructure policies is to deal with this fundamental contradiction. The future concept of strategic development must therefore reconcile the provision of privately owned logistics platforms – required by the companies involved – with long term and rigid public infrastructure investment programmes. Cost/benefit assessments might be necessary. Subsidies by governments are justified in the first phases of such developments.

Logistics platforms, hubs, teleports are part of the response that modern economies require in terms of accrued performance from companies and prompt reaction to market developments and needed structural adjustments taking into account the great diversity of demand. They reorganise physical flows through an extension and intensification of consolidation/degrouping operations and a reduced number of storage points. This was made possible and efficient through information systems and technologies which allowed the information corresponding to these more numerous and more complex physical movements to circulate reliably; large operators are all equipped with very modern information processing systems benefitting from the rapid progress made in this field.

In short, the typical logistics operation scheme for "platforms" is coupled with a specialised information and communication infrastructure. The great number of actors operating in the logistics chain impose the need for standardisation and interoperabilitly of systems such as Cargo Community Systems (CCS). In spite of the difficulty for EDI to penetrate the transport sector, this information and telecommunications system is considered to have a high potential for future development. In the meantime, it is interesting to note the extension of the use of the Internet by transport operators. The Internet can be considered as a complementary tool to EDI and its extensive use in the future will probably lead to new economic and industrial organisational structures compelling telecommunication providers to remove some institutional barriers in a more competitive context.

II.3.4. *Assessing key needs and developments*

In order to confront ideas, discussions were held between researchers, administrators, shippers, operators, managers and practitioners at two international OECD.RTR workshops:

- The International Workshop on *Integrated advanced logistics and information technology in freight transport*, organised in February 1994 by the US Federal Highway Administration in Washington D.C.;

- The International Workshop on *Integrated advanced logistics and innovations in freight transport*, organised in March 1995 by the Japanese Ministry of Construction in Tokyo.

These OECD Workshops and the two OECD reports provide a series of conclusions developed hereunder.

Business globalisation

The implications of business globalisation and the integrated complex – production - product marketing - logistics strategies – are quite radical in regard to both organisational structures and IT applications. Better understanding of the dynamics of business and the spatial development of investment patterns, particularly with respect to logistics, will therefore help all parties involved to anticipate future needs in policy and decision making.

Innovations

Innovations occur at a rapid pace in physical networks, information systems, management practices and inter-entreprise co-ordination. Innovations in logistics affect business operations in all sectors as regards productivity, job creation and competitiveness. Innovation takes place in all parts of the world and occurs in different countries, at different times and in different ways.

In a number of cases, public partners and/or administrations can play a critical role in co-operating with industry. Developing standards for deployment of new technologies is a typical case. Continuous efforts to highlight and publicise innovations are beneficial. Pilot projects which demonstrate the value of innovations to industry are commended.

Advanced logistics and information highways

Advanced logistics are critically interconnected with information technologies and information highways. Integrated logistics are not possible without innovations in information technology as an integrating tool. To clarify the overall structure and relationship between freight transport and information highway networks, it is important to define the systems architecture and the structure of information processing attached to the logistics operations. This underscores the importance of standards in logistics and explains why the architecture can be an important building block in the development of a set of standards.

Role of governments: Challenge or necessity?

Governments can have a significant impact by promoting the use of information technologies throughout the logistics sectors. Governments are responsible for the planning, investment, pricing and decisions of the transport and telematics infrastructure, which supports the provision of logistics services. The development of logistics affects environment, health and safety, which are important concerns of governments. In importing and exporting there are many possibilities for innovation in government procedures which reduce the time and costs of inspection. Governments are also required to provide support to SME's on adopting EDI and computerisation so as to maintain their viability and survival. Particularly, inter-governmental actions to reduce standardisation and regulation barriers are important in the world-wide globalisation of business.

Governments also have an important role in monitoring the security of new logistics systems and if required providing legislation and implementing systems that will ensure fair competition between all sectors involved.

Removing the barriers through co-operation

As discussed earlier, the major innovations in information technology and organisation design require new forms of collaboration. There are new levels of integration in the restructured business processes. Several important examples show the effectiveness of collaborative actions, such as the development of Cargo Community Systems (CCS) or Value Added Network (VANS) to provide information interchange among multiple trading partners.

As far as the VANS are concerned, the Port of Rotterdam and the Schipol Airport are operated respectively through the NITS and Cargonaut system (marine-ground and air-ground interfaces). Similarly, in Singapore, seventeen government agencies have simplified both import and export clearances for trade in creating the Trade Net System.

In North America, the major railroads and private car fleet owners are collectively organised through the Association of American Railroads in order to deploy the Automatic Equipment Identification (AEI) System, using activated radio tags. Such multi-party processes and multi-organisation collaboration are requirements in many cases, but they are also difficult to implement.

II.4. TRILOG PROJECT

II.4.1. *Logistics in response to global trade*

Globalisation invariably means an increase in the international movement of goods. The increased dependence on intermodal transport and the new types of information systems in support of logistics strategies imply a number of challenges to governments.

Logistics is characterised by a diversity of players with different rules, different economic interests, different cultures and different market conditions. Different strategies will work for different players and there is a value in learning from the diversity. While the most developed countries may lead in some areas, other countries in other regions are also innovating in different ways.

In order to monitor these changes and their impacts on future transport policies, the Programme has launched a global project on advanced logistics focusing on a trilateral regional approach comprising Europe, NAFTA countries and the Asian-Pacific. The TRILOG Project was launched in 1996 with a three-year timeframe.

II.4.2. *Charting the future course*

The aim of the TRILOG Project is to stimulate the exchange of information about logistics and information technology on a multi-regional basis. Amongst the objectives to be met by TRILOG, one of the key question to be answered by the project is: What could be the role of governments to capture full benefits through advanced logistics as a contribution to sustained economic growth? And what should be the balanced role of road transport and the trucking sector?

The project is being implemented through three major regional Task Force Groups: Europe, North America and Asia-Pacific. The European and North American regions face major challenges in advancing effective integrated advanced logistics and freight transport systems. As far as the Asia-Pacific region is concerned, there is a need to assess the current status of and bottlenecks to logistics and freight transport in the various Asian countries and their connections to the other world regions.

TRILOG takes full account of the main achievements of the OECD Seminar on *Prioritisation of multimodal infrastructure* (see Section II.2.1) as well as the results of the Conference on *Intermodal networks and logistics* (Section II.2.2).

The first step of TRILOG

At the initiative of Japan, the Asia-Pacific Task Force organised a planning meeting of TRILOG in Singapore in April 1996 in order to set up the main orientations of the project. It was recognised that the continuing growth of global business interconnections and international trade had created the conditions for further progress in the logistics sector. Key issues include: the potential of recent IT/communications technologies (such as ITS systems) taking into account the social need for a high performance information infrastructure (information highways) in order to realise efficient, safe and friendly transport systems – advancement of logistics in the information society: more diversified consumers' needs due to the large volume of information and access to user-friendly information systems -- the recognition of the need for regional analyses based on global interests, concepts or common frameworks that assure a global platform for the trilateral study – network development characterised by the intermodal aspects of globalised logistics i.e. the co-ordination, prioritisation and financing of modes – the role of governments and the need for standards and common architecture to ensure openness and interoperability of information systems which are vital for efficiently operating globalised logistics systems.

The second step of TRILOG

Jointly organised with the OECD Symposium on *Globalisation and advanced logistics* held in November 1996 in Fukuoka (Japan), the first plenary meeting of TRILOG took place under the chairmanship of Japan. Based on the contents of national and regional reports, the regional task forces will work along three directions:

- Assessment of the problems and opportunities at local and regional levels;

- How the strategy developed in a region is impacting on the strategies developed in other regions, from a logistics standpoint;

- Integration of regional strategies at global level: needs, opportunities and requirements.

48

As far as the research methods are concerned, liaison between the various actors involved in TRILOG will be ensured by the APEC Transportation Sub-group, the EU/Directorate General VII and the North American Committees.

The next step of TRILOG

In the framework of the 1997 Mexico Conference on *Intermodal transport networks and logistics*, the TRILOG Group will evaluate the results of the Seminar on *Prioritisation of multimodal transport infrastructure*, including the conclusions drawn from the questionnaire replies on the experience of OECD Member countries on intermodality and intermodal systems developed and processed by the Dutch Delegation. Emphasis will be on recommendations for applications as they emerge from the presentations on *Innovative financing of transport infrastructure from a modal and intermodal approach*. A NAFTA focused OECD Seminar is planned to be held in Toronto in October 1997.

II.5. REFERENCES

1. OECD. (1996). Seminar on *Prioritisation of multimodal transport infrastructure*. The Netherlands.

2. OECD. ROAD TRANSPORT RESEARCH (1992). *Advanced logistics and road freight transport*. OECD. Paris.

3. OECD. ROAD TRANSPORT RESEARCH (1994). *Congestion control and demand management*. OCDE. Paris.

4. OECD. ROAD TRANSPORT RESEARCH (1996). *Integrated advanced logistics for freight transport*. OECD. Paris.

5. OECD. Conference on *Intermodal networks and logistics* to be held on 3-5 June 1997 in Mexico.

6. OECD ROAD TRANSPORT RESEARCH and VTI. (1987). *Just-in-time transport: New road freight transport strategies and management: Adapting to the new requirements of transport services, Part II*. Research Seminar arranged by OECD and VTI, 22-24 June 1987, in Svenska Mässan, Gothenburg (Sweden).

CHAPTER III: ADVANCED TRAFFIC TECHNOLOGIES AND PRACTICES

III.1. TRAFFIC GROWTH OVERWHELMS ROAD DEVELOPMENT

Automobile ownership, road travel, and freight shipment by road have been increasing at astounding rates. The growth of road traffic is brought about by either the entry of new users or as a result of a shift from another mode, such as rail, to road. In a 1994 publication entitled *Congestion Control and Demand Management* (1), the OECD estimated that urban travel alone will experience a fifty percent increase by the year 2005. It also estimated that this increase in travel will bring about an increase in the delays experienced by road users. Delay times could increase by as much as 400 percent in the same time frame.

The importance of the figures on the increase of travel and delay is that several major problems accompany them. The most obvious problems are economic. For every minute that an automobile or truck is delayed there are associated costs related to, among other things, the value of the driver's time, the use of fuel, and the costs associated with not getting a product to its destination as quickly as possible. In addition, safety and environment are both adversely affected by increased traffic.

Though increasing travel and congestion are recognised as critical issues in all OECD Member countries, as a general rule road investments lag behind what would be necessary to properly address infrastructure limitations. However, even if sufficient funding were available, it is likely that the problems would still exist because the conventional approach of building more roads is hampered for political, financial, social, and environmental reasons. In addition, there is more and more resistance to building new roads because it is believed that it has proven to often compound the problem by simply inducing a more rapid rate of travel growth.

The challenge for OECD and Non-OECD countries alike is, therefore, to identify or develop the ways and means to alleviate traffic-related problems without building new roads. The two principal ways in which these problems can be addressed without new roads is through the application of better traffic management measures and the development of new technologies.

III.2. DEVELOPING INTERNATIONAL EXPERTISE THROUGH RESEARCH CO-OPERATION

III.2.1. Overview of activities

Early in its history, the Programme recognised both the common needs of Member countries in the area of traffic research and co-ordinated research conducted by individual countries. Thus, the Programme concluded the first of many co-operative international traffic research activities when it published

Table III.1. **OECD Traffic Research Publications**

YEAR	TITLE
1971	Electronic Aids for Freeway Operation
1972	Area Traffic Control Systems
1975	Research on Traffic Corridor Control
1978	Integrated Urban Traffic Management
1979	Traffic Measurement Methods for Urban and Suburban Areas
1981	Traffic Control in Saturated Conditions
1982	Traffic Control and Driver Communication (Seminar)
1982	Automobile Fuel Consumption in Actual Traffic Conditions
1983	Traffic Capacity of Major Routes
1984	Micro-Electronics for Road and Traffic Management (Seminar)
1985	Energy Savings and Road Traffic Management
1987	Dynamic Traffic Management in Urban and Suburban Road Systems
1988	Route Guidance and In-Car Communication Systems
1991	Future Road Transport Systems and Infrastructure in Urban Areas (Seminar)
1990/1992	Road Tunnel Management (Seminar)
1992	Intelligent Vehicle Highway Systems: Review of Field Trials
1994	Advanced Road Transport Technologies (Seminar)
1994	Congestion Control and Demand Management

Electronic Aids for Freeway Operations in 1971. Table III.1 lists the year and title of traffic-related documents that have been published by OECD/Road Transport Research Programme.

Table III.1 illustrates several things. It illustrates the fact that problems related to increasing traffic are a continuing high priority for OECD countries. It also illustrates that measures and technologies that are put in place to address traffic related problems are in a constant state of review and improvement. It illustrates that as long as these problems persist, governments must work to find ways to alleviate the adverse affects associated with them. Most tellingly, however, the table reflects the continuing role the Programme has played from its nascence by convening Scientific Expert Groups and organising seminars around this very important subject. It therefore underscores the value that OECD has to offer to its Member countries by creating a forum for them to share experiences and find common, workable solutions.

III.2.2. OECD/RTR and ITS

One of the traffic management areas that has received a high level of attention in recent years is Intelligent Transport Systems (ITS) formerly known as Intelligent Vehicle Highway Systems (IVHS). These advanced technologies are beginning to have an impact on the scope and direction of international co-operative programs. ITS technologies are being developed and implemented to solve existing transport problems for which traditional solutions are no longer feasible. Such is the case with traffic management and congestion relief applications.

ITS applications are used to create efficiency where current practices impose important financial and time-related costs for users. Sample applications in this case are heavy vehicle and commercial driver electronic data systems, weigh-in-motion technologies, and automated toll collection facilities. ITS technologies are also making the roads friendlier and safer for average users through the application of technologies such as route guidance systems, incident response systems, and accident warning systems. Finally, ITS research in various countries and regions is exploring new and varied applications that can potentially revolutionise transport systems throughout the world. The Programme is paying close attention to these developments and staying involved with key countries and regions, primarily OECD Members, involved in ITS development.

The Programme completed its most recent ITS-related Scientific Expert Group with the publication of *Intelligent Vehicle Highway Systems: Review of Field Trials* (2). The report captured the expert opinions from the group following a review of IVHS functions, systems, test area characteristics, field trials, policy and implementation issues, and the conclusions and recommendations resulting from field trials throughout the world. Immediately following the completion of this report, the RTR Steering Committee instructed the Secretariat to begin development of a Seminar related to IVHS technologies in order to create an avenue for more sharing of research and implementation experiments and results.

In June 1994, the OECD, in co-operation with the Japanese Ministry of Construction and the National Police Agency of Japan, hosted the *Seminar on Advanced Road Transport Technologies* (3). It was attended by more than 400 people and included 25 presentations by experts from the OECD and Member countries. The objective of the Seminar was to highlight the problems associated with the growth of motorisation when road development does not keep pace and to suggest possible solutions through the application of ITS technologies. Based on the presentations and discussions, the organisers identified the conclusions and recommendations listed in Box III.1.

Though many of the recommendations resulting from the seminar rely upon the individual action of the Member countries for follow-up, some provide avenues for future RTR Programme involvement. In particular, the last recommendation makes a strong call for activities that are suited to the OECD keeping in mind that they must policy oriented or research specific so as to complement the triade ITS World Congresses starting in Paris (1994) and then in Yokohama (1995) and Orlando (1996). Therefore, the Programme Steering Committee continues to discuss co-operative activities in this technical area and to seek out policy research needs that could profit from OECD assessment during the upcoming triennial period.

III.3. TRAFFIC MEASURES THAT BRING RESULTS

When confronted with the problems of explosive traffic growth coupled with the physical, political, environmental or financial constraints on expanding road infrastructure, road managers are inclined to search for non-traditional solutions. ITS technologies for automobiles and trucks, as described in section

Box III.1. Ten Recomendations on Advanced Road Transport Technologies (Omiya, 1994)

1.   Advanced road transport technologies have been developed as a means to solve traffic congestion, environmental issues and other concerns in OECD Member countries.

2.   Provision of new infrastructure is difficult in many countries and has limits. There is no single traffic measure that can solve the problem of congestion. Therefore, comprehensive transport plans are necessary. Advanced transport technologies can mitigate congestion, but at the same time incentives and financial measures acting on road users will have to be considered.

3.   Great expectations are placed on advanced road transport technologies as a way of controlling traffic congestion in order to facilitate more efficient road use and thus to solve environmental issues attributable to road traffic as well as to improve traffic safety. To achieve these aims, further research and development is required.

4.   Advanced road transport technologies have diverse applications, for example in traffic demand management, dynamic route guidance and traffic flow control. However, careful examination is required to ensure optimum application for maximum effect, and application strategies also need to be prepared. Integration of technologies/systems and interoperability should be systematically considered.

5.   The effects of advanced road transport technologies on traffic safety is not yet fully understood. Further study is required on how to use these technologies to improve traffic safety.

6.   For advanced road transport technologies to enjoy widespread adoption, it is necessary to identify who will benefit and how. It is also necessary to identify user expectations and find out how they evaluate these technologies. In addition, technologies that are easy to use need to be offered. Further research on these subjects is necessary.

7.   It is necessary to address the problem of cost allocation for advanced road transport technology deployment between end users, industries and authorities.

8.   Proposals suggested during the seminar included new transport systems such as vehicle sharing, using electric vehicles and a new freight transport system. These systems could be a substitute for automobiles and help to reduce vehicle traffic, ease congestion and protect the environment. Research and development on these new systems are required. There is also a need to look into ways to make public transport more attractive and thus to adjust traffic demand.

9.   Full political support and commitment in the application/implementation of advanced technologies is essential.

10.   It was recognised throughout the seminar that it is very important for experts of OECD Member countries to exchange knowledge and experience and promote international co-operative research for advanced road transport technologies. It is strongly recommend that co-operative research in this field be positively pursued.

III.2 and the on-going section, offer some solutions. However, as with infrastructure and logistics strategies for truck traffic, ITS technologies are not a panacea for the universal problem of traffic congestion. The road manager is called upon to consider as many alternative options as possible when dealing with the problem.

In this spirit, the OECD convened a Scientific Expert Group on *Congestion Control and Demand Management* (1). Chaired by the United States, the Expert Group began its efforts by clearly stating the problem. Taking the road traveller's point of view, the OECD study ascertained that traffic delays were growing in all participating OECD countries. From an employer's point of view, congestion takes a toll in lost worker productivity, delivery delays, and other direct costs. The Group described the road traffic congestion problem as not being solely a problem confined to commuter trips in large cities or urban areas. Rather, they asserted that congestion adversely affects the non-work trip as well. It affects the movement of people and the flow of goods. In non-urban, or rural areas and inter-city corridors, traffic is disrupted by incidents, maintenance operations, detours, over-loaded tourist routes, and other causes.

The final report presents a collection of measures under the heading *"Congestion Management"*. The measures are considered to fall under two classifications: demand-side and supply-side. Demand-side congestion management measures are those designed to reduce car dependency and car demands on the road system. Supply-side congestion management measures are intended to increase the existing capacity of the system in order to improve traffic flow for all modes. Implemented individually or in concert with one another, congestion management measures can help to achieve one or more objectives listed in Box III.2.

Box III.2. Congestion Management Objectives

1. Reduce the need to make a trip.
2. Reduce the length of a trip.
3. Promote non-motorised transport.
4. Promote the use of public transport.
5. Promote carpooling.
6. Shift peak-hour travel.
7. Shift travel from congested locations.
8. Reduce traffic delays.

Table III.2. **Congestion Management Measures**

TYPE OF MEASURE	STRATEGY CLASS	MEASURES
DEMAND-SIDE	LAND-USE AND ZONING	Land-use and Zoning Policy Site Amenities and Design
	COMMUNICATIONS SUBSTITUTES	Telecommuting Tele-Conferencing Tele-Shopping
	TRAVELLER INFORMATION SERVICES	Pre-Trip Travel Information Regional Rideshare Matching
	ECONOMIC MEASURES	Congestion Pricing Parking Pricing Transportation Allowances Transit and Rideshare Financial Incentives Public Transport Pass Programmes Innovative Financing
	ADMINISTRATIVE MEASURES	Transportation Partnerships Trip Reduction Ordinances and Regulation Alternative Work Schedules Auto Restricted Zones Parking Management
SUPPLY-SIDE	ROAD TRAFFIC OPERATION	Entrance Ramp Controls Traveller Information Systems Traffic Signalisation Improvements Motorway Traffic Management Incident Management Traffic Control at Construction Sites
	PREFERENTIAL TREATMENT	Bus Lanes Carpool Lanes Bicycle and Pedestrian Facilities Traffic Signal Pre-emption
	PUBLIC TRANSPORT OPERATIONS	Express Bus Services Park and Ride Facilities Service Improvements Public Transport Image High Capacity Public Transport Vehicles
	FREIGHT MOVEMENTS	Urban Inter-city

The Expert Group developed a "Catalogue of Congestion Management Measures" that can assist road managers to achieve these objectives. 35 conventional and innovative measures are presented within nine strategy classes. Table III.2 shows the measures and how they fall into strategy classes and whether they are demand-side or supply-side measures.

In addition, the report presents an overview of the policies, plans and programmes implemented in different countries and cities. To emphasise actual experience, the report presents case studies that illustrate organisational details, operations and impacts. Finally, the report considers current and future societal policies on the environment, economy, and advanced technologies in order to present a picture of how congestion management measures could be developed and shaped in order to be more effective.

The conclusions and recommendations that resulted from the work of the Expert Group are summarised below. They were presented by the group in such a manner as to allow them to be considered as policy advice regarding the application of congestion management measures in OECD countries.

- Road traffic congestion can be better managed.
- Low-cost, conventional measures can be effective.
- Pricing techniques can be effective in congestion relief.
- Public Support of congestion management measures is essential.
- Traveller information is important to congestion relief.
- Co-ordination is an essential aspect of congestion management.
- Congestion management efforts need to start small then grow.
- The private sector has a role in congestion management.
- New policies and laws are needed for congestion management.
- New technologies will offer tools for congestion management.
- Accessibility must be maintained with congestion management.
- Evaluations are needed in congestion management.
- Training in the practices of congestion management is needed.

Taken in their entirety, the recommendations present strong, direct and yet simple advice on why congestion management is necessary and how a road administration can begin to implement cost effective measures immediately. In total, the report therefore serves as a guide for OECD Member and non-member countries alike on the how these measures can be adopted and implemented in order to bring about real and quantifiable benefits to road users.

III.4. CONTINUING TO MAKE PROGRESS

The problem of road traffic congestion is severe and potentially getting worse. Individual countries have labored in their research and development programmes to develop new methods and technologies that can ameliorate this problem. In addition, the Programme has been called upon numerous times to foster international co-operation among Member country traffic experts to help advance the state-of-the-art and practice in this important area.

In spite of these efforts, more needs to be done. From this standpoint, the OECD is prepared to continue to serve the Member countries by doing its part in seeking better and more widely adopted solutions to this common problem. It is forseeable that technological measures alone will not suffice. Even with considerable resistance and many reservations at the policy level, economic measures – such as road pricing – might have a greater chance for large scale application as we begin the next century.

III.5. REFERENCES

1. OECD. ROAD TRANSPORT RESEARCH (1994). *Congestion Control and Demand Management*. OECD, Paris.

2. OECD. ROAD TRANSPORT RESEARCH (1992). *Intelligent Vehicle Highway Systems: Review of Field Trials*. OECD, Paris.

3. OECD. ROAD TRANSPORT RESEARCH (1994). *Proceedings: Seminar on Advanced Road Transport Technologies* held in Omiya (Japan). OECD, Paris.

CHAPTER IV: INFRASTRUCTURE

A country's road and bridge network is an enormous national asset. Management of the system is a highly sensitive and complex task, entrusted to the road administration and shaped by a constellation of political, technical, environmental, managerial and historical forces. Changing environmental concerns, social structures and the scale of road networks has increased the necessity for ever greater optimisation in terms of cost, use of materials and management.

The political environment in which the road administration manager works is complex and sometimes even considered hostile by some professionals. Managers are asked not only to attempt to minimise the society's expenditure on transport but also to meet user needs. These user needs range from an acceptable level of service, to a desire for a clean environment, sustained economic development, and a low share of user taxes to provide for the agency's funding. The everyday reality for road administration professionals is that they must increase their effectiveness and productivity in accomplishing a mission that grows in complexity while budgetary constraints become increasingly severe. The difference between user costs and agency costs is striking and must be kept in mind (see section IV.1).

IV.1. TOP PRIORITY ON ROAD REHABILITATION AND MANAGEMENT

The emphasis is on road maintenance and rehabilitation, although in many OECD countries, new construction is still ongoing to fill in the missing links. The OECD programme has a long tradition of co-operation in road maintenance research. The study on *Pavement Management Systems* (1) reviewed the status of systems in use at the time and proposed a procedure to develop the principal building blocks. The report recognised that even the best of the systems at that time left room for improvement.

Figure IV.1. **Cost-shares under optimal maintenance**

50 veh./day

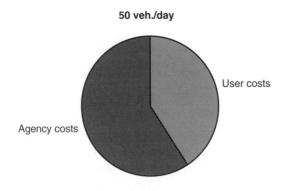

300 veh./day

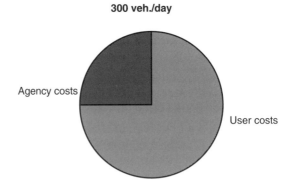

5 000 veh./day

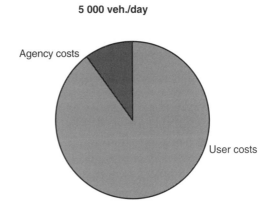

One of the most striking and serious omissions was the apparent failure of the systems to specify a quantifiable statement of goals and objectives that compared the positive and negative impacts of pavement conditions, intervention levels and techniques on all concerned parties – highway authorities, users and the community at large. Narrow technical objectives, such as the need to prevent an index of overall pavement quality from falling below "x", were too often described without examining their full consequences. There was an obvious lack in quantifying the impacts of pavement management decisions on user costs.

In sum, the first generation of pavement management systems seemed to be designed to achieve goals of the responsible engineers and administrators which were not necessarily identical with optimising the functioning of pavements for users and the community at large. Since then two OECD studies have been completed addressing these issues: *Road Maintenance and Rehabilitation: Funding and Allocation Strategies* (2) – developing a systematic programme for use by road administrations to allocate funding – and *Performance Indicators for the Road Sector* (3) – to assist road administrations and governments evaluate their performance in providing a cost efficient service and identify areas for improvement and cost cutting.

IV.1.1. Funding and allocation

The asset value of the road network is estimated at 1.5 to 3 times the Gross National Product (GNP) of a country and the costs of transport in the economy ranges from 2 percent to as high as 17 percent for less developed countries of the Gross Domestic Product (GDP). The value of this asset and the need to maintain its functionality coupled with both an ageing road network and increased traffic loading has inevitably meant increased investment on maintenance and rehabilitation.

The study on *Road Maintenance and Rehabilitation: Funding and Allocation Strategies* (2) undertaken in co-operation with the World Bank, developed a framework for minimising user and administration costs and a system for allocating often limited funding where it was most effective and necessary. The report identified "Ten Commandments" to be observed by governments and road administrations for resource allocation and distribution for road maintenance and rehabilitation programmes. These "commandments" can be seen in Box IV.1.

Box IV.1. **Resource allocation for road maintenance and rehabilitation programmes**

1. Maintenance is an opportunity for enhancing the environment as well as safeguarding the road network asset.

2. Road and bridge maintenance should be pursued for the sake of the users. Therefore, public participation is an essential part of developing the road maintenance programme.

3. Road and bridge assets should maintained in an economical way.

4. A sound analytical framework is important for delivering an economical and environmentally sound product.

5. User costs must be treated as important costs and included in the analytical framework.

6. Budget constraint on the administration's expenditure is an important feature of the analytical procedures. Competitive maintenance and rehabilitation programmes are one important means to address these constraints.

7. The entire road budget and trade-offs between alternative uses must be considered when allocating and distributing resources.

8. The management systems used in allocating and distributing resources must be compatible with the road administrations organisation and management style.

9. The methods used at the network, programme and project levels must be different but interlocking and utilise the same database.

10. Data systems which support the road and bridge management systems must be timely and reliable.

IV.1.2. *Strategic approach to performance measures*

Public administrations are increasingly faced with scrutiny and accountability. Historically, demands are placed on their efficiency to deliver services to the public at minimum cost, but increasingly administrations must meet annual service level targets, especially in the intermodal policy context of today. Road administrations are no exceptions to these demands. In reality, many, if not most, road administrations operate in a complex political environment and are themselves complex organisations with multidimensional outputs whose quantity and quality are difficult to measure. Against this background, the OECD has recognised the need to develop, jointly and internationally, a set of standardised indices, *performance indicators,* which road administrations can use to gauge themselves.

Many road administrations have already developed performance indicators to evaluate their results. These indicators often are, however, 'ad hoc', lack peer review and are difficult to translate across different management environments. So far, their usefulness to road administrations has therefore been limited.

International co-operation in developing and testing performance indicators can be helpful in enhancing the state of the art and in assisting road administration management to be responsive to decision-makers and to users. In response, the Programme established a Scientific Expert Group on *Performance Indicators for the Road Sector* (3) to perform a study in co-operation with the World Bank.

The Group identified and classified performance indicators in 8 categories: Accessibility and Mobility, Safety, Environment, Equity, Community, Programme Development, Programme Delivery, and Programme Performance. Each category has 3 perspectives: Government level, Road administration level, Road user level. 40 primary indicators and 35 secondary indicators were identified. Of the primary indicators, 16 key measures are detailed in Table IV.1 and were considered by participating OECD countries to form a minimum common set. To test their validity and practicality in a real world context, an international OECD Task Force has been established to carry out a field test using these 16 key indicators.

Performance indicators can serve in 6 broad circumstances: (i) examination of issues for the development of alternative courses of actions and performance targets; (ii) evaluation of the achievement of goals and objectives; (iii) assessment of the efficiency and effectiveness of the road administration; (iv) management of road administrations; (v) development or periodic re-evaluation of goals and objectives; and (vi) as an aid to a learning organisation.

Performance appraisal can not be conducted in the absence of reasonably accurate and continuous data systems. Data systems developed to support performance appraisals also provide a rich source of system performance and financial analysis data for system evaluation and programme and budget development. Adequate appraisals can be done with generally available data supplemented by additional sources.

Table IV.1. **Key indicators for the international field test**

PERSPECTIVE DIMENSION	GOVERNMENT (Ministry)	ROAD ADMINISTRATION	ROAD USER
ACCESSIBILITY MOBILITY	• Average road user cost (car and truck)		• Level of satisfaction regarding travel time, its reliability and quality of road user information
SAFETY	• Accident risk: Fatalities and injury accidents/veh.-km (and the number of fatalities and injured)		• Unprotected road user risk
ENVIRONMENT		• Environmental policy/programme (y/n)	
COMMUNITY		• Processes in place for market research and customer feedback (y/n)	
PROGRAMME DEVELOPMENT	• Long term programmes for construction, maintenance and operations (y/n)	• Management systems for distribution of all the resources(y/n) • Quality management/audit programme (y/n)	
PROGRAMME DELIVERY		• Forecast values of road costs vs. the actual costs (%) • Overhead percentage	
PROGRAMME PERFORMANCE	• Value of assets	• Roughness • Percentage of defective bridge deck area	• Surface condition • Satisfaction with road condition

IV.2. SEARCHING FOR INNOVATIVE TECHNOLOGIES

IV.2.1. DIVINE

Apart from climatic effects, heavy vehicles such as trucks, cause by far the largest proportion of highway wear. A co-ordinated highway-truck transport policy is needed to ensure maintenance costs are minimised. A study into the *Dynamic Loading of Pavements* (4) concluded that suspension standards, axle weights and axle configurations could be set to both reduce road wear and increase vehicle productivity. Infrastructure could support these objectives by being designed to minimise and resist moving dynamic wheel loads. This could also be taken into account when reconstructing and rehabilitating roads.

There is, however, insufficient scientific evidence to quantify implementation measures. In particular, on the highway side, the magnitude of the effect of moving dynamic wheel loading needs to be quantified. On the vehicle side, there is a lack of convergence of vehicle tests to assess their effects on pavement distress. Aside from the economic benefits of optimisation, the rapid growth in heavy vehicle traffic in all countries, increasing wear of pavements and bridges, and increasing innovation in vehicle design provide both pressure and opportunity to develop a scientific basis for dealing with dynamic road loading and road wear.

The project

Taking these considerations into account, a major international investigation has recently been completed by the OECD, focusing on the dynamic interaction between vehicle suspensions, pavements and bridges. The project consisted of six interrelated research modules which have produced new findings. The aim was to provide government agencies with sound scientific evidence on which to base their policies, and supply vehicle manufacturers with invaluable information to assist them in the design optimisation of suspensions.

The project known as DIVINE (*Dynamic Interaction of Vehicle and Infrastructure Experiment*) involved specialists from seventeen countries in the areas of vehicles, pavements, bridges, road management and transport policy. The inter-linked research projects were carried out in nine countries. The total international budget amounted to 1.5 million U.S. dollars and was financed by participating OECD countries, industry and the European Commission. Figure IV.2 presents the organisational structure.

Element 1: Accelerated pavement test

The accelerated dynamic loading test undertaken in New Zealand compared the performance of two different suspension types, air and steel. A circular full scale pavement was specially constructed and instrumented and subjected to 1,700,000 loading cycles over ten months. The outer wheel path of the track

Figure IV.2. **Management and organisation structure of the DIVINE Programme**

Executive
Chair: Denmark
France   Netherlands
United
Australia   United

Scientific Expert Group
Chair: Australia
Participating countries:
Australia, Austria, Belgium, Canada,
Denmark, Finland, France,
Hungary, Japan, Netherlands,
Japan, Netherlands, New Zealand,
Sweden, Switzerland,
United Kingdom, United States

Programme
United Kingdom

ELEMENT 1
Research Leader: New Zealand
Test site: New Zealand
Equipment: CAPTIF Facility
Research team: Australia, Finland,
France, Germany, New Zealand,
Sweden, United States

ELEMENT 4
Research Leader: Netherlands
Test: Evaluation of selected
computer packages, Netherlands
Research team: Germany,
Netherlands, New Zealand,
United Kingdom

ELEMENT 2
Research Leader: United States
Test sites: Finland,
United Kingdom, United States
Equipment: Trucks from Canada,
United Kingdom, United States
Research team: Canada, Finland,
United Kingdom, United States

ELEMENT 5
Research Leader: France
Test sites: France, United Kingdom
Equipment: trucks from Canada,
United Kingdom, Germany
Research team: Canada, Finland,
France, Germany, United Kingdom

ELEMENT 3
Research Leader: Canada
Test site: Canada
Equipment: Canadian shaker;
Trucks from Canada,
United Kingdom, United States
Research team: Canada,
New Zealand, Sweden, Switzerland,
United Kingdom, United States
Data analysis: Canada,
Germany, Hungary

ELEMENT 6
Research Leader: Switzerland
Test sites: Switzerland and
Equipment: Trucks from
Canada, Australia
Research team: Australia, Canada,
France, Switzerland, United

was trafficked using a steel suspension and the inner wheel track using air suspension. It has been confirmed and quantified that deteriorating road profiles cause increased dynamic loading for the steel suspension compared to the air suspension. The extensive data collected throughout this test can be used to improve pavement performance models.

Element 2: Primary pavement dynamic response testing

Pavement response measurements to live loading were examined in the United States, Finland and the United Kingdom. Pavements were instrumented and tested using instrumented vehicles. Pavement responses, for example the response due to a long wavelength bump in the road, were measured under controlled conditions. The analysis quantified the differences between suspension types in terms of frequency, amplitude and damping. It is anticipated that this pavement response information will deepen the understanding of deterioration mechanisms and help develop and improve pavement design methods.

Element 3: Road simulator testing

Vehicle dynamics were also researched to assess vehicle road friendliness. A database of measurements of dynamic wheel forces and dynamic wheel behaviour under both on-road and laboratory road simulator conditions was established. The DIVINE research undertaken in Canada has provided insights into the differences between laboratory and on-road conditions and the situations under which the laboratory tests are likely to produce the best results. It has been possible to accurately reproduce on-road conditions in the laboratory and the availability of a validated library of such vehicle parameters will be of great use in computer and other simulation models. Methods of assessing vehicle suspensions for their effects on pavements have been clarified.

Element 4: Comparison of computer simulation of vehicle dynamics

The data collected from instrumented vehicles tested on-road and in the laboratories were used to compare existing simulation models and to test them for accuracy. The study was undertaken by the Netherlands. The availability of validated useable models will assist in policy implementation. It has been found that some models can reproduce dynamic wheel loads with a certain amount of precision, however the input parameters have to be detailed and correct if the results are to be accurate. Details of the suspension system, spring and dampers are critical to modelling vertical responses. Performance was quite widely varied among the models tested.

Element 5: Measuring the spatial repeatability of dynamic loads

Longitudinal loading profiles on roadways have shown that certain areas on roads are subjected to higher and more frequent dynamic loading than others. This phenomenon, known as spatial repeatability was investigated in France. In design it has always been assumed that loads are randomly distributed along the road surface. The occurrence of spatial repeatability could have an impact on the future design of roads. The occurrence of spatial repeatability was studied on two distinct roads using varying speeds, vehicles, suspensions and conditions. Surface roughness was found to play an important part in spatial repeatability, however the phenomenon was also apparent on smooth roads to a lesser extent.

Element 6: Dynamic loading of bridges

The response of bridges to traffic loading was studied in Switzerland and Australia. In particular, the study examined the main causes for dynamic responses thereby showing the principal requirements for bridge friendly suspensions. The frequency of the structure, profile of the road and the dynamic responses of the vehicles all inter-relate and act together. There are many factors aside from vehicle suspensions which affect the dynamic response of bridges. These include: pavement roughness; bridge length and hence natural frequency; bridge materials; profile of approach road; and, in particular, a bump at the start of the bridge causes increased dynamic loading. Damping of both the structure itself as well as the vehicle suspension is one of the most important components to reduce the extent of dynamic responses.

Policy and outputs

The individual research institutes have published the research elements in separate technical reports and the OECD will compile these into a CD-ROM. A *Final Technical Report* (5) summarising and referring to the element reports will be published by the OECD. In addition, a policy document will be produced to draw on the conclusions of the study in order to assist in the formulation of future policy actions for governments. Three conferences are planned: one in Canada aimed at North American countries; one in The Netherlands for the European region; and one in Australia directed towards Australasian countries. They are intended to disseminate the research findings and tailor the policy implication discussions to the particular regions. Different emphases are required in these regions because of varying legislation and design standards regarding both infrastructure and vehicles.

IV.2.2. *Recycling system for road improvements*

The report entitled *Use of Waste Materials and By-Products in Road Construction* (6) examined the environmental and economic aspects of recycling certain waste and by-product materials as an acceptable solution to the problem of resource depletion. Given the great number of changes that have come about in this field after the 1977 report, the OECD established a Scientific Expert Group that produced a new report on *Recycling for Road Improvements* (7) in 1997.

The purpose of the group -- chaired by the United States -- was to assess and report on current practices and policies related to the engineering and environmental aspects of recycling for road construction and rehabilitation. Ultimately it is believed that the recommendations made in this effort will promote more recycling of waste and by-product materials. The environmental implications associated with this report are covered in Chapter VI, Environment/ Transport Interaction, while the engineering elements are discussed here.

The report produced by the Scientific Expert Group acknowledges that most Member countries have made significant advances in their understanding of the technical performance of recycled materials used in road construction. The Group surveyed Member countries on the use of road and non-road by-products in road improvements and the report provides the results of this survey which identifies several workable technologies (winners). The discussion of the technical qualities of the materials is concluded with a detailed presentation of those by-products that can be considered to have the greatest potential for recycled use in road projects.

Governmental laws, policies and regulations can be restrictive so as to reduce the production of waste, control disposal and limit the use of new materials. Governmental regulations can also be promotional and subsidise recycled by-products, research, testing, evaluation and demonstration of recycled materials. The report makes it clear that a balance of regulations and policies is paramount to the success of a recycling programme.

The development of adequate information on the long-term performance of by-products, standard specifications and testing requirements will continue to slow the recycling movement. The study team encourages: an increase in

IV.5. REFERENCES

1. OECD. ROAD TRANSPORT RESEARCH (1987). *Pavement Management Systems*. OECD. Paris.

2. OECD. ROAD TRANSPORT RESEARCH (1994). *Road Maintenance and Rehabilitation: Funding and Allocation Strategies*. OECD. Paris.

3. OECD. ROAD TRANSPORT RESEARCH (1997). *Performance Indicators for the Road Sector*. OECD. Paris.

4. OECD. ROAD TRANSPORT RESEARCH (1993). *Dynamic Loading of Pavements*. OECD. Paris.

5. OECD. ROAD TRANSPORT RESEARCH (1997). DIVINE (Dynamic Interaction of Vehicle and Infrastructure Experiment) Final Technical Report. OECD. Paris.

6. OECD. ROAD TRANSPORT RESEARCH (1977). *Use of Waste Materials and By-Products in Road Construction*. OECD. Paris.

7. OECD. ROAD TRANSPORT RESEARCH (1997). *Recycling System for Road Improvements*. OECD. Paris.

8. OECD. ROAD TRANSPORT RESEARCH (1976). *Bridge Inspection*. OECD. Paris.

9. OECD. ROAD TRANSPORT RESEARCH (1992). *Bridge Management*. OECD. Paris.

10. OECD. ROAD TRANSPORT RESEARCH (1995). *Repairing Bridge Substructures*. OECD. Paris.

CHAPTER V: ROAD SAFETY

V.1. CONTINUING POLICY AND RESEARCH CHALLENGE

Judging from facts and figures over the last twenty-five years, road safety policy has been a success story in most OECD countries. Figure V.1 depicts past trends since the peak in the seventies (1). Some OECD countries report accident levels in 1995 as low as 1956, and this in spite of increasing traffic. Overall, the risk of being fatally injured in a traffic accident has declined.

Both the public interventions by governments and private sector initiatives, foremost by the automobile industry, have been the driving forces that pushed effective safety performance in developed countries. In no small measure, research and scientific analysis have been a significant contributing factor. Clearly, road safety is an area where policy and research are closely interrelated. No agenda for safety programmes is complete without an agenda for R&D.

Today, more precise estimates of the traffic accident situation world-wide have been obtained. The series of OECD/RTR conferences and seminars in CEEC's/NIS (1994, 1995) (2), Asia (1993, 1996), Africa (1997) and Latin America (1995)(3-6) have provided more reliable data. By and large, there are annually over 500,000 road traffic deaths and 15 million injuries world-wide, i.e., 1,400 fatalities and 41,000 injuries per day – of which respectively about one quarter and one third, respectively, are in OECD countries. This compares to 22.6 millions AIDS infected persons world-wide with a total death count of 1.5 million up to the present; in 1996 the increase was 3.1 million infections, i.e., 8,500 per day.

These numbers show that traffic safety remains an important public and social priority issue as well as an economic challenge, since the economic losses amount to 1-2 per cent of GDP in OECD countries or up to 4 per cent if estimates are based on the "willingness to pay" methodology.

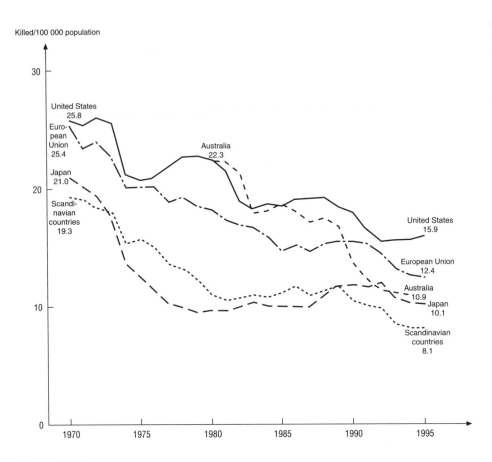

Figure V.1. **Trends in fatalities per 100,000 population in selected OECD regions**

Killed/100 000 population

Source: IRTAD.

Figure V.2. **Trends in fatalities per 100,000 population in selected countries**

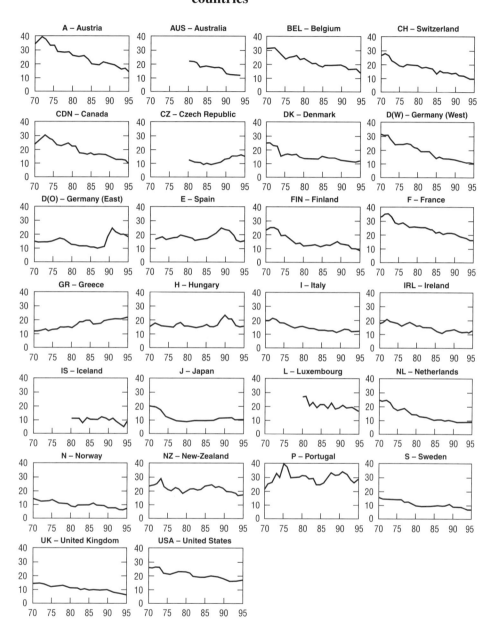

Source: IRTAD.

V.2. ROAD SAFETY RECORDS: PAST, CURRENT AND FUTURE

Figures V.1-V.3 present the overall picture of the traffic accident situation and developments in OECD regions and countries based on fatalities per 100,000 population. The statistics were drawn from the Programme's *International Road Traffic and Accident Database* created in 1990 (see Chapter VII.2).

It can be seen that fatal accident risks vary considerably between OECD Member countries ranging from 6.5 killed per 100,000 inhabitants in some countries to almost 30 in others. The lowest risk rates reported were in Sweden, Norway, the UK, Finland and The Netherlands; the highest rates reported were in Portugal, Greece and (former) East Germany. West Germany, Switzerland, Denmark, Ireland and Japan hold a medium position. All other OECD Member countries report medium to high risk rates.

In several OECD countries, there is now solid evidence for stagnating or again somewhat increasing fatality, injury, and especially injury accident trends linked partly to certain (high risk) road user groups. Therefore, new initiatives are needed. Safety policy-makers are required to make tough decisions, because the safety measures likely to be introduced will be socially and publically pervasive and need long-term commitments and investments, especially in regard to highway infrastructure.

What does the future hold? Predictions of future safety developments have been proposed by SWOV, the Dutch Road Safety Research Foundation, [see OPPE (7 and elsewhere), KOORNSTRA (3,6) and WEGMAN (21)] based on time-series analyses and other methodologies. It is worthwhile to quote from the OECD Summary report *"Road Infrastructure Rehabilitation and Safety Strategies in Central and East Europe"*(2) which is specifically related to CEEC's where traffic accident risks are dramatically high and motorisation has been increasing at an unprecedented pace:

> "The growth of motorisation is accompanied by an exponentially decreasing curve for fatality rates. Just by combining both developments as a product [fatalities = fatalities/kilometrage × kilometrage] the development of fatalities could be described. This leads to the conclusion that a reduction in the number of fatalities ought to be the result of a higher decrease in fatality rate than an increase in mobility growth. *A reduction rate of 8-10 per cent in fata-*

Figure V.3. Country comparison of fatalities per 100,000 population in 1995

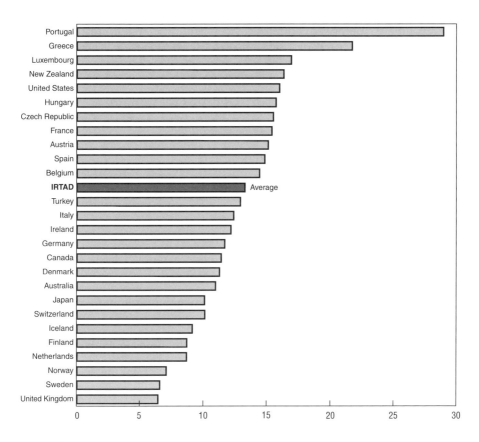

Source: IRTAD.

lity rates must be considered realistic targets for Central and Eastern European countries. If traffic growth is not accompanied by appropriate risk reducing countermeasures and activities, an increase of road fatalities might be the outcome. The lesson to be learned from highly motorised countries is, when accelerated traffic growth is anticipated, **no time can be lost to invest in safety**."

V.3. GOALS-ORIENTED STRATEGIES

V.3.1. *Defining force of road safety targets*

Most OECD countries have committed themselves to trying to reduce the number of traffic casualities by setting targets within fixed time frames. It is recognised (8) that:

- Targets and targeted planning provide a rational basis for a national consensus on priorities and for regional and local initiatives as well;

- Setting targets leads to realistic and coherent programmes with a wider scope;

- Setting targets ensures better use of public funds and other resources;

- Targeted planning gives direction to policy making and encourages commitment of all partners;

- Comprehensive monitoring schemes must accompany these programmes to evaluate results and update them.

The 1994 OECD report on *Targeted Safety Programmes* (8) found that the time frame of quantified targets was typically either 3-5 years or about 10 years. The level of ambition different from one country to the other and ranged from a target of preventing an increase in the number of deaths and injuries to reducing their number by a total of 45 per cent or by 2 to 5 per cent per year. To shed some light on the actual difficulties, the statement by Mrs. T. Linnainmaa, Minister of Transport and Communications of Finland, at the 1995 Helsinki IRTAD Seminar (7) is worth quoting:

> "By the year 2000, we aim to reduce the number of persons killed in traffic accidents to a level that equals half of the total number of persons killed in accidents in 1989. By now, we are within the target, but hard efforts are needed in order to achieve the final goal in 2000.

This will be even more difficult as we know that the vehicle kilometrage will grow".

As pointed out above, the annual numbers of fatalities are strongly influenced by traffic levels, with past data showing a linear relationship and suggesting that road safety programmes in many countries have succeeded in holding constant the number of fatalities against a traffic growth of about 4-5 per cent per year. Target setters should therefore relate their targets to explicit assumptions about traffic growth.

Non-controversial and easily implementable accident countermeasures are lacking. Political, budgetary and societal imperatives and influences often override purely economic or technical solutions. It is no wonder therefore that the most common organisational problem reported in regard to targeted programmes is the lack of an integrated process that assigns responsibility for implementation. A critical challenge is how to link actual capabilities and resources of a safety agency to national safety goals and the strategy chosen. Action depends largely on the motivation of each organisation involved to advance with measures falling within its own area of responsibility. Co-ordinators need considerable skill to reconcile different viewpoints and interests.

V.3.2. *Reconciling safety vs. environment*

Ideally, safety and environment strategies should be mutually reinforcing. To study issues and opportunities, a Scientific Expert Group on *Integrated safety/environment strategies* was created in 1995 and will report in 1997(10).

In most OECD countries the cost of congestion and the cost of accidents amount each to about two per cent of GDP, with noise and local air pollution costing a further 0.5 to 1 per cent of GDP. Estimates of the long-term cost of CO_2 emissions to global warming vary between one and ten per cent of GDP (9).

There is a need for techniques that achieve satisfying trade-offs between conflicting or non-converging aims. In many cases, parallel but separate strategies for planning road safety and environmental protection – especially in urban areas or along the major transit corridors – are no longer warranted. Notwithstanding, separate strategies acting towards parallel aims can result in overall improvement even if no specific co-ordination has taken place. To the extent that, for instance, the objectives of safety and environment strategies overlap regarding health concerns, the measures implementing these strategies can reinforce each other.

The long term solution is to embed the integration of road safety and environmental aspects in the broad concept of sustainable development which stipulates stringent demands on the transport system as a whole. For the shorter term, the prime need is to ensure the productive interaction of environmental and safety strategies, while seeking a balance within the large scope of transport policy. If such a balance can be achieved, some directions for the development of integration can be seen in:

- Using overall regulatory and economic instruments to achieve complex objectives;

- Adopting a transport corridor approach and seeking a holistic solution;

- Focusing on a strategic approach, especially in urban areas, ensuring acceptable mobility, while taking environment and safety into account.

The report (10) provides a detailed discussion of the potential points of convergence and conflict. A limited number of case studies are reviewed. The instruments available in order to achieve road safety and environmental protection objectives are identified (see also Chapter VI). There is a need for developing specific, operational methods of assessing how transport schemes and projects implement the strategic objectives set.

Public involvement in environment/safety programmes is strongly recommended. This requires transparency of the assessment as well as simplicity and clarity in the presentation of plans and their impacts.

V.4. MOTIVATING ROAD USERS AND MARKETING SAFETY

The potential and limits of attitudinal research in underpinning innovative road safety initiatives were investigated by the OECD Safety Research Group on *Improving Road Safety by Attitude Modification*(11). This subject is a serious matter and not a playing field for free-thinking newcomers to road safety policy. Fifteen experienced traffic psychologists came together to scrutinise the state of the art in this complex problem area.

The report shows how motivating attitude change can play a useful role in improving road safety. The "producer" - the road safety agency - must understand the attitudes as well as the motives of the "consumer", the road user. Still, attitudes of road users cannot be analysed properly without taking account of political, cultural, economic and technological factors and influences. On the one hand, these add to the complexity of behavioural controls and attitude modification; on the other hand, there are ways to use these influences more constructively.

It was found that specific and operational information and messages focusing on attitude modification will be effective. A case in point is the application of employer-based programmes which can constructively use the company's influence to prescribe employees to use seat belts or to avoid speeding. Also, the teaching and introduction of road safety principles from an early age in homes, kindergartens and schools will be beneficial and long lasting.

Similarly, a new mind-set was envisioned by launching an international review on *marketing of traffic safety*(12). It was found that social marketing can be effectively used as an instrument in road safety programmes and as a method of reinforcing conventional approaches, e.g.,

- in the preparation of regulatory measures;
- as a support of safety initiatives by private insurers;
- in assisting school traffic education;
- in implementing road safety publicity measures.

The OECD marketing concept has been very well received by safety professionals. It was presented at several international meetings in 1994-1996 and promoted especially by private sector associations and the automobile industry. There is now ample potential to further develop, promote and apply marketing in accident prevention and road safety work. Pilot evaluation projects could be useful for particular case studies of major interest, novel safety measures and selected target groups such as young/new drivers, school and/or small children, elderly people, drunk drivers, professional groups such as policemen, certain groups of employers, etc.

Marketing should be applied as a strategic principle not solely when comprehensive programmes are planned, but also – and perhaps especially – when only modest financial resources are available.

The *application of social marketing* in road safety contains distinct elements, many of which are new in product promotion: *(i)* it focuses much more on the target road user and his/her perceptions and motivation; *(ii)* it requires the road safety expert, practitioner or policy maker to adapt set objectives to the perceived requirements, costs and benefits of the target group; *(iii)* it encourages promotion of remedial measures and communication with the public on the anticipated benefits and socio-economic costs to be incurred.

A *marketing plan* begins with an analysis of the situation one wishes to improve. Once the market has been segmented and the target group identified, objectives are set and the marketing strategies to achieve them are formulated. These strategies often require a sensitive balance of marketing elements - product, price, promotion and place - before an effective social/marketing initiative can be undertaken. And after its implementation, monitoring and evaluation can feed in information on results that will guide future road safety initiatives. At all points of this process, indeed, market and consumer research is facilitating the exchange of information and thus allowing substantial improvements in effectiveness and acceptance of safety measures.

V.5. PRIORITY ON VULNERABLE ROAD USERS

The safety of vulnerable road users is under renewed scrutiny in OECD countries. This is in large measure due to the commitment of community leaders to help and protect particular social and age groups – children, the elderly, the disabled – which are especially exposed to road traffic risks as pedestrians and two-wheelers. There is also a political dimension because the growing trend is to redirect traditional transport policies towards a greater share of non-motorised traffic that is less polluting and energy consuming than automobiles.

An extensive international survey and research review was therefore launched in 1995 to revisit this problem area in the light of new information, expertise and needs. The final OECD report on *Safety of Vulnerable Road Users* is expected to be issued in 1997 (13). It updates earlier OECD work undertaken in the 1970's and 1980's which was the basis for a series of ECMT recommendations to Ministers.

The modal share of fatalities in 1993 in some selected OECD countries is shown in Figure V.4. As can be seen, vulnerable road users have a share of

22 percent (United States) to 60 percent (Japan). Developing countries all have shares exceeding 50 percent. In spite of the safety improvements achieved so far in absolute numbers, large scale progress is still possible to reduce the frequency and severity of accidents to vulnerable road users, while facilitating their mobility.

Figure V.4. **Modal share of fatalities in 1993 in some selected countries**

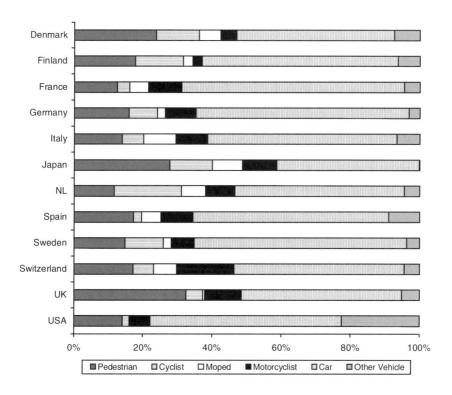

The OECD report provides a detailed analysis of the accident situation and past developments as well as highlighting the relative differences between OECD countries. It is difficult to generalise, but very broadly speaking there is increasing concern for the elderly (65+), the lack of safety of moped riders (and motorcyclists), the age groups 15-24, or under 14, depending on countries, as well as cyclists. Again, much of this is a function of the national context.

Due to the underreporting of such accidents and the non-availability of exposure data, definite statistical assessments – especially at the international level – cannot be expected in the near future. Not to mention the mobility/safety problems of disabled persons for whom firm data is lacking.

What can be done? Undoubtedly, many safety measures taken in the past in favour of vulnerable road users have been effective and successful: dedicated infrastructure and networks; traffic calming; improving visibility and conspicuousness; protective devices, education, etc. The OECD group calls for farsighted programmes and creative solutions. Its recommendations centre on the following areas:

- Change in attitudes and social climate; higher priority in favour of this user group;

- Gradual adaptation of the legal environment (statute, rules, regulations);

- Further development of innovative facilities, both inside and outside built-up areas as well as in transition zones; intersection treatment;

- Better communication and information;

- New technology in accident prevention and protection.

A series of guiding principles is put forward. It is intended to disseminate the extensive information gathered and to report to the competent international regulatory and policy bodies for follow-up. It is also foreseen to initiate further contacts with developing countries' experts, especially in Asia, who have great experience – and problems – with two-wheeled traffic.

V.6. TRUCK DRIVING AND FREIGHT TRANSPORT

Trucking and road transport has been a strategic research focus of the Programme for many years. This issue with all its different facets is a chronic component of any government's transport policy or their attempts to balance costs and benefits in the infrastructure sector. Governments are determined to see an improvement in the economic performance of operators and truckers' safety record. This aim is in principle supported by the road transport industry and road transport workers as a whole.

During the period under review, two priority safety areas were singled out for in-depth study:

- qualifications, licensing and training of truck drivers, and
- transport of dangerous goods through road tunnels.

V.6.1.   *Training truck drivers*

Heavy freight vehicles are involved in a higher proportion of fatal accidents than other vehicles, even though they are under-represented in all traffic accidents. Rough estimates indicate that heavy freight vehicle accidents account for at least 15 per cent of total road accident costs.

The OECD report on *Training Truck Drivers* (14) reviews human factors, working conditions and driver fatigue; industry structures and corporate culture; licence acquisition practices and regulations; driver training programmes and new (simulation) technologies; and institutional aspects.

Surveys on truck drivers' conditions show that they spend more time on the job than workers in other sectors. A working week of 50 to 60 hours is common, and drivers are often away from home for more than three consecutive days. The long hours on duty and modern traffic requirements impose demanding working conditions. This creates serious demands on driver performance and causes fatigue. The problem is now under intense investigation in many OECD countries and will demand careful consideration and appropriate actions. This is a burning issue as highlighted so dramatically by the French truckers' strike in November 1996.

With the deregulation of the trucking industry, limiting costs becomes extremely important in the face of the new competitive forces motor carriers face every day. For these reasons, training imposed by regulatory control is often viewed as a hindrance and an additional cost. There also seems to be a relationship between, on the one hand, the size of the trucking company, and on the other hand, its willingness to implement training programmes.

The largest road freight transport companies, who do require high standards for truck drivers, find a clear disparity between the level of training they require and the level of training demanded by the licensing process. The companies that have undertaken extensive driver training identify lower accident rates and lower fleet operating costs – again strengthening the argument for training beyond the minimum legal requirement for licensing.

The industry is made up of thousands of small firms within thin profit margins which makes it difficult to invest in training. But they might be responsive to fiscal and financial incentives. Governments should not hesitate to introduce market-based measures and explore ways to reward good safety records through fiscal incentives.

Among ways to more rapidly develop driver experience after initial training, governments could consider graduated licensing requiring several years on smaller trucks before moving to larger equipment. Also the emergence of new technologies, such as simulators, has offered new scope for safety training, driving skills and experience enhancement including fuel-efficient driving. The costs for these simulators – US$ 1.5-2 million – are quite substantial and their shared financing and co-ordinated use in a country or region are recommended. R&D should be pushed in this field.

The area of truck driver licensing and training is changing, in a context of political and economic change in Europe as a whole, the Americas and Asia/Pacific. The whole area of training for trucking companies may change in the future, with the need for transport companies to focus on innovative business logistics and to use ITS technologies. This need will be driven by both shippers and large transport organisations.

V.6.2. *Reducing risks in tunnel operation*

In the quest for safer tunnel operation and protection against hazardous environmental spills, a Scientific Expert Group on the *Transport of Dangerous Goods Through Road Tunnels* was established in 1995 to conduct joint international research. The activity initiated by the United Kingdom and France is a follow-up to the 1988 OECD report on *Transport of Hazardous Goods by Road,* (15) the 1990 Seminar on *Road Tunnel Management* held in Lugano and the 1992 Seminar on *Strategies for Transporting Dangerous Goods by Road: Safety and Environmental Protection (17),* held in Karlstad, Sweden, in co-operation with the OECD Environment Directorate.

There are more and more tunnels on the road networks of Member countries, more traffic in tunnels and more transport of hazardous materials. While such transport – because of the precautions taken – is generally "safer" than normal truck transport, the accidents that did occur in Los Alfaques (Spain), in the Nihonzaka Tunnel on the Tomei Expressway in Japan, or in Herborn (Germany), had catastrophic consequences in terms of human and economic losses. The recent Channel tunnel accident in November 1996 is another warning, although railway tunnels are in principle easier to control than road tunnels.

Authorising or prohibiting the transport of dangerous materials in certain road tunnels can have important implications on the cost of transport, and indeed the organisation of the companies producing or using these materials. The research project should identify the most suitable provisions for reducing the risks from the point of view of tunnel design, construction or equipment as well as maintenance, operating procedures, new ITS technologies and the organisation of emergency response. The implementation of clear, applicable and coherent regulations from one tunnel to another and from one country to another, will be beneficial for rational management of such transport, and therefore for economic development.

It is essential to have methods which can ascertain whether it is less dangerous or more dangerous to authorise the transport of hazardous substances in a given road tunnel rather than use alternative routes above ground.

The project is divided into four tasks:

1. *Review of current national and international regulations*

   To provide an overview of current regulations, with an evaluation of the requirement and controls imposed, and the problems they pose; to draw lessons on directions and needs as well as new regulations.

2. *Quantitative risk assessment and decision support models (QRA and DSM)*

   To recommend a general quantitative assessment methodology and propose a standard model approach for evaluating risks in tunnels.

3. *Risk reduction measures (including transport and tunnel operations)*

   To recommend measures well adapted to each specific case, with detailed specifications and evaluation of costs and benefits.

4. *Conclusions and recommendations*

   To propose a standard formulation for international and national regulations for authorising or prohibiting the transport of dangerous goods, including application provisions; to recommend measures to reduce risks.

Progress has been substantial in 1996. A Project Director from Austria was designated in October. The first phase of Task 1 was completed by the analysis of 50 questionnaire replies received from OECD countries and tunnel operators world-wide. There were striking differences and in-depth interviews, are now programmed by the Norwegian consultant. Any recommendations will be built on the existing ECE agreements for the international carriage of dangerous goods by road for the road transport of dangerous goods. Task 2 was launched through an international Seminar on *Risk Assessment and Decision Making Process* in Oslo in March 1996 with the participation of forty high-level experts under the chairmanship of Denmark. An international call for offers to develop QRA and DS models was issued in December 1996.

So far nine countries (Austria, Denmark, France, Japan, Norway, the Netherlands, Sweden, Switzerland and the UK) are funding the project with voluntary grants. Representatives of three countries – Belgium, Italy and the United States – are participating through the Tunnel Committee of the Permanent International Association of Road Congresses (PIARC). The Commission of the European Communities is currently considering financial support to the proposed overall budget of US$ 1.2 million, although the Commission's preferred focus would be the surveillance of railway tunnels. Once full financing is assured, the joint research project could be completed by the end of 1997, as initially envisaged. The final report and recommendations should be approved and available in 1998.

V.7. RESEARCH PERSPECTIVES

A critical overall theory for road safety is lacking. If future policies and actions are to be fully successful, a comprehensive theoretical basis for road safety is needed. This is clearly recognised by the research community active within the OECD Road Transport Research Programme. To assess what exists, a Scientific Expert Group on *Models in Road Safety* was established in 1995. The final report will be available in 1997 (19).

A review of past developments reveals that road safety work has mainly been practical in nature. Indeed, the need for theories, models and research methodologies originated from practice, where the value of systematic know-

Table V.1. **Evolution of road safety paradigms (19)**

ASPECTS	PARADIGM I	PARADIGM II	PARADIGM III	PARADIGM IV
Decennia of dominating position	1900 - 1925/35	1925/35 - 1965/70	1965/70 - 1980/85	1980/85 - present
Description	Control of motorised carriage	Mastering traffic situations	Managing traffic system	Managing transport system
Main idea and FOCUS	Use CARS as horse drawn carriages	Adapt people to manage traffic SITUATIONS	Eliminate risk factors from road traffic SYSTEM	Consider exposure of risks, regulate TRANSPORT
Motor vehicles / 1000 pop.	Less than 25	25 - 250	250-500	>500
Main disciplines involved	Law enforcement	Car and road engineering, psychology	Traffic engineering, traffic medicine, advanced statistics	Advanced technology, systems analysis, sociology, communications
Organisation of vehicle production	Craft-production, craftsmen's manufacturing	Mass-production workers assembling	Lean production, group assembly on sub-contracting	Recycling materials
Terms used about unwanted events	Collision	Accident	Crash, casualty	Suffering, costs
Role of persons using motor vehicles	Ownership of vehicles: "Car owner"	User of motor power: "Motorist"	Active part of the system: "Driver"	Social partnership: "Road-user"
Attitudes towards automobiles	Fearful curiosity	Blind admiration	Prudent tolerance	Calm consideration
Premise concerning un-safety	Transitional problem, passing stage of maladjustment	Individual problem, inadequate moral and skills	Defective traffic system	Risk exposure
Data ideals in research	Basic statistics, answers on "What"	Causes of accidents; "Why"	Cost/benefit ratio of means "How"	Multidimensional
Organisational form of safety work	Separate efforts on trial and error basis	Co-ordinated efforts on voluntary basis	Programmed efforts, authorised politically	Decentralisation, local management
Typical countermeasures	Vehicle codes and inspection, school patrols	The three E's doctrine, screening of accident prone drivers	Combined samples of measures for diminishing risks	Networking and pricing
Effects	Gradual increase in traffic risks and health risks	Rapid increase of health risk with decreasing traffic risk	Successive cycles of decrease of health risks and traffic risks	Continuous reduction of serious road accidents

ledge is increasingly recognised. However, it is only relatively recently that systematic research based on theories and models has reached a recognised status in road safety work.

The OECD Group starts with the following message in the first chapter: *There is nothing as practical as a good theory.* Indeed, a theory or a model is not only for the researcher. They are very practical in the research process and in actual road safety work – and also in the interaction between researchers, between practitioners and between researchers and practitioners.

Historical perspectives can open a vision for the future. Table V.1 presents an overview of the past evolution, and describes possible developments for the future. Road safety research is now moving into an increasingly scientific and cross-disciplinary phase. This reflects the steadily growing complexity and density of the road traffic process, affecting the typology of safety problems.

To systemise the discussion of theoretical approaches and models so far applied, the OECD Group identified five principal types of conceptual research approaches and models: descriptive, explanatory-aggregated, explanatory-individual, accident consequences, implementation focus. While generally speaking, statistical and econometric approaches as well as risk factor modelling and, with some limitations, biomechanical models are operational and have proven effective for well defined safety problems, the fifth road safety approach – pointing to implementation – is very much needed, but generally missing. How can existing empirical and theoretical knowledge be made useful for, and transferred into practical road safety work?

Research concerning implementation problems and related modelling approaches is a key challenge for future success in road safety work. There is an abundance of road safety research results available, both at national and international level. Most of these results, however, are not fully used, as practitioners cannot ascertain appropriate applications. It is unfortunate but this problem rarely attracts the interest of researchers.

Following this line that stresses implementation, two further priority recommendations are put forward:

- Emphasis on result management rather than activity management as the main thrust of road safety management systems;

- Creation of an independent evaluation and auditing system for road safety work.

These three topics – implementation modelling, road safety management systems, and investigative auditing – could be amongst the priorities for future international co-operation.

V.8. REFERENCES

1. OECD. ROAD TRANSPORT RESEARCH. International Road Traffic and Accident Database (IRTAD) (1996). *Accident Statistics*. OECD/BAST. Paris/Bergisch-Gladbach.

2. OECD. ROAD TRANSPORT RESEARCH (1995). *Road Infrastructure Rehabilitation and Safety Strategies in Central and East Europe*. OECD. Paris.

3. OECD/MALAYSIAN GOVERNMENT (1993). Proceedings of the Conference on Asian Road Safety 1993 (CARS '93). Kuala Lumpur.

4. OECD/CHINA SOCIETY OF TRAFFIC ENGINEERING (1996). Proceedings of The Second Conference on Asian Road Safety (CARS '96). Beijing.

5. ECA/OECD. Third African Road Safety Congress. 14-17 April 1997, Pretoria.

6. OECD/PIH (1995). International Conference on *Traffic Safety for Latin American and Caribbean Countries*. São Paulo.

7. OECD Seminar on *International Road Traffic and Accident Databases*, (1995). Helsinki.

8. OECD. ROAD TRANSPORT RESEARCH (1994). *Targeted Road Safety Programmes*. OECD. Paris.

9. OECD/ECMT (1995). *Urban Travel and Sustainable Development*. OECD. Paris.

10. OECD. ROAD TRANSPORT RESEARCH. Scientific Expert Group on *Integrated Safety/Environment Strategies*. To be published in 1997. OECD. Paris.

11. OECD. ROAD TRANSPORT RESEARCH (1994). *Improving Road Safety by Attitude Modification.* OECD. Paris.

12. OECD. ROAD TRANSPORT RESEARCH (1993). *Marketing of Traffic Safety.* OECD. Paris.

13. OECD. ROAD TRANSPORT RESEARCH. Scientific Expert Group on *Safety of Vulnerable Road Users.* (to be published in 1997). OECD. Paris.

14. OECD. ROAD TRANSPORT RESEARCH (1996). *Training of Truck Drivers.* OECD. Paris.

15. OECD. ROAD TRANSPORT RESEARCH (1988). *Transporting Hazardous Goods by Road.* OECD. Paris.

16. OECD/Swiss Federal Highways Office (1992). *Road Tunnel Management.* Final Report of the OECD Seminar held in November 1990 in Lugano. OFR. Berne.

17. OECD. Seminar on *"Strategies For Transporting Dangerous Goods by Road: Safety and Environmental Protection".* (1992). OECD Environment Monograph no. 66, June 1992. Karlstad, Sweden.

18. OECD. ROAD TRANSPORT RESEARCH (1996). Transport of Dangerous Goods through Road Tunnels -- Risk Assessment and Decision-making Process: Methodologies, Models, Tools. Seminar Proceedings. Oslo, March 1996. OECD/PIARC, Paris/Oslo.

19. OECD. ROAD TRANSPORT RESEARCH. *Models in Road Safety* (to be published in 1997). OECD. Paris.

CHAPTER VI: ENVIRONMENT/TRANSPORT INTERACTION

VI.1. REVISITING ENVIRONMENTAL ISSUES

Environmental policy has made significant strides forward over the last twenty-five years. In the 90's the environmental impacts of transport infrastructure, traffic demand and mobility requirements have again come under close scrutiny in OECD Member countries. The complexity and nature of environmental issues have underscored the need for a world wide approach. This led to the organisation of a number of international Conferences and concerted actions with programmatic declarations and white policy papers such as the Rio Agenda 21 of 1992.

In most OECD countries, the largest share of transport activity is by road. Road transport is responsible for over 80 per cent of final energy consumption for transport. Over the last forty years there has been a ten fold increase in the number of motorised vehicle in the world to over 700 million today. For the forthcoming decades, it is expected that both the number of vehicles and the amount of travel will grow substantially. Overall activity involving heavy vehicles will increase even more.

Table VI.1. **Estimated trends in road transport**

	Light vehicles			**Heavy vehicles**		
	1990	2030	variation (%)	1990	2030	variation (%)
Number of vehicles (millions)	468	811	73	16	31	94
Kilometres travelled (billion)	7,057	12,448	76	687	1,377	100

Source: OECD/Environment Directorate (1)

This Chapter presents the main outcome of key research undertaken by the OECD/RTR since its 25th Anniversary in 1993 (see Box VI.1). The Programme had pioneered road related environmental research in the early 70's. With the creation of the OECD Environment (Policy) Committee, the OECD's work was centralised and an impressive amount of high level reviews, joint study reports, statistics and guidelines was produced by the OECD Environment Directorate. Consensus on a number of policy directions in numerous environmental fields was achieved with emphasis on the relationship between economic and environmental policies.

Box VI.1. 1994-1997 OECD/RTR Projects in the field of Environment/Transport interactions

1994	Report on *Environmental Impact Assessment of Roads*
1994	Seminar on *Environmental Impact Assessment of Roads - Strategic and Integrated Approach*, Palermo, June 1994 (co-sponsored by EU/DGVII)
1995	Workshop on *Environmental Impact Evaluation of Road Infrastructure*, Prague, February 1995 (co-sponsored by EU PHARE)
1995	Report on *Roadside Noise Abatement*
1995	Seminar on *Roadside Noise Abatement*, Madrid, November 1995
1995-97	Scientific Expert Group on *Integrated Safety/ Environment Strategies* (Report to be published in 1997)
1995-97	Scientific Expert Group on *Recycling for Road Improvements* (Report to be published in 1997)
1995-98	Research Project on *Transport of Dangerous Goods through Road Tunnels* (co-operation with PIARC Tunnel Committee)

Responding to new initiatives by RTR Member institutions, the RTR Programme strengthened its activity in the 90's focusing on transport/environment interaction research and selective road environment issues. The main goal of the Programme was to identify and evaluate measures that reduce the impact of roads on the environment, from a technological and scientific point of view. Efforts centred on environmental impact assessment, noise reduction, transport of hazardous goods in road tunnels, integration of safety and environment aspects, recycling and the potential offered by innovative transport logistics.

VI.2. KEY ELEMENT IN TRANSPORT POLICY

VI.2.1. *Environmental impact assessment*

The set of principles drawn up at the Rio Conference and the new widely accepted idea of "sustainable development" spurred renewed interest in environmental protection and in ways and means of establishing a link between assessment and public communication. The report *on Environmental Impact Assessment of Roads* (2) was issued in 1994 to review the planning process in OECD countries and the strategic component governing environmental requirements. The study is based on a comprehensive survey in OECD countries and on selected case studies. It provides a review of traditional environmental assessment methods used at *"project" level* and procedures currently used in the road and road transport sector. It explores the potential of new research developments focusing on *strategic approaches* for long-term policies, plans and programmes.

As a follow-up to the Group's conclusions and recommendations, a Seminar (3) was held on the same topic in 1994 with the aim to develop a series of Recommendations, which are summarised in the *10 recommendations of Monte Pellegrino* (Box VI.2). These have been widely disseminated and the principles have been accepted by transport and environmental planners.

Based on the experience of OECD Member countries and in co-operation with the OECD Environment Directorate, the Road Transport Research Programme organised in February 1995 in Prague (Czech Republic) a Workshop on *Environmental Impact Evaluation of Road Infrastructure* in the framework of the series of 14 Workshops towards CEECs and NIS (4) (see Chapter VIII). The Workshop identified key issues of environmental impact evaluation in relation to road infrastructure and road transport development in CEECs and NIS. It is hoped to follow up this initiative in 1997.

Box VI.2. The 10 Recommendations Of Monte Pellegrino

I.   OECD Member countries should examine the scope for integrating Strategic Environmental Impact Assessment (SEIA) into the overall assessment of transport policies and plans at various levels (e.g. conurbation, regional, national, international). All transport modes and vehicles must be taken into account in order to improve the balance and complementarity between modes and achieve sound and sustainable mobility.

II.   Strategic Environmental Impact Assessment (SEIA), for both national land use planning and transport management, should be co-ordinated as much as possible and should serve as a guide and a reference to Environmental Impact Assessment of projects.

III.   Environment must be considered by decision makers as important a factor as mobility, accessibility, safety and economics. The use of well-defined goals and policies is essential for formulating alternative actions. New emphasis and new methods are needed to actively seek better environmental conditions, instead of only avoiding or mitigating environmental damage.

IV.   Efficient co-ordination between the administrations responsible is essential for the success of the work carried out by road agencies. Decisions by transport authorities should be based on a joint process shared with the other agencies involved, they must work together to determine purpose and needs, the scope of impact assessment and the environmental mitigation and improvement measures.

V.   SEIA and EIA must integrate cultural, social and natural background. Road administrators must consider social values and keep in mind new ethics, such as those resulting from the Rio Agreement: the precautionary principle and the ability of future generations to meet their own needs.

VI.   SEIA and EIA methodologies should be part of the educational curriculum of road administrators, in order to enhance their expertise in project and policy formulation.

VII.   Short, medium and long term effects must be considered. SEIA and EIA should not be single actions but a continuous, long-lasting process. The responsibility of road designers does not end with the project, and the decisions resulting from SEIA and EIA need monitoring, follow-up and review.

VIII.   Participation by the public and the other parties of the decision process require involvement at an early stage of the process. Consequently, project leaders need to be trained in the techniques of communication.

IX.   The knowledge about short, medium and long term effects of roads on the environment must be enhanced. Research on monetary and non-monetary valuation should be developed. National policy must determine how these approaches are used in the decision-making process.

X.   The highway administrations of Member countries should undertake case studies to examine the possibilities for improving SEIA and EIA methods, and then share their experience. The harmonisation of SEIA and EIA methodologies is necessary especially for international projects. However, constraints of ensuring conformity which could undermine the efficiency of national actions should be avoided. The process of harmonisation and the development of a common language should continue through ongoing international mechanisms.

VI.2.2. Integrated environment/ safety strategies

The impacts of a transport system are diverse and affect both and simultaneously the environment and the safety of road users and roadside residents. Both aspects should be weighed equally in transport and urban and regional planning (see also section V.3.2). This is the objective of a Scientific Expert Group of the OECD (5) which started its work in 1995 with the aim of assessing the interface between the safety and environmental aspects of road transport. The Group undertook to prepare a comprehensive review of evaluation methods and planning tools for both aspects, in view of possibly proposing guidelines for integrated strategies.

Clearly the quality of the interaction between environment and safety objectives is essential to a successful strategy. This is typical for urban areas, where the greatest concentration of problems is encountered. In the past, public authorities have generally given priority to reducing congestion. As a result, environmental and road safety concerns were given a secondary priority. The public is now becoming more and more aware of the consequences of transport on its welfare and demands more attention on both environment and safety.

The approach towards environment and safety aspects has been generally fragmented. There are several explanations: the different focus of safety and environmental objectives, different actors involved, responsibility divided between different levels of governments.

As pointed out in Section V.3.2, separate uncoordinated strategies acting towards parallel aims may result in overall improvement but they may also have unfortunate results. A typical example concerns the implementation of a noise barrier, for which there are often conflicts arising regarding noise attenuation, landscape impacts and traffic safety. One can also mention the use of de-icing salts for road winter maintenance and safe winter driving; this may have negative consequences on the environment. To advise East European countries on newest technologies and winter service standards a Workshop on *Road Winter Maintenance* was held in Prague (Czech Republic) in October 1994 in the framework of the series of 14 Workshops towards CEECs and NIS (4).

The problems of road safety and environmental protection have in common that they are tackled by governments by means of public policies which will influence the behaviour of consumers. Problems should be considered very early in the planning process. In general it is difficult to resolve all the conflicts particularly those between mobility and safety, mobility and the environment,

Table VI.2. Instruments available in order to achieve road safety and environmental protection

Instruments	Impacts			
	Accident	**Noise**	**Air pollution**	**Energy/CO2**
1. Regulations				
a. *Vehicle standards*				
• emissions (pollutants, noise)		X	X	
• size/weight/power	X			
• energy efficiency				X
• active/passive safety	X			
b. *Town and country planning standards*				
• density, zoning		X	X	
• construction		X		
• parking			X	
c. *Infrastructure standards*				
• safety improvements	X			
• noise		X		
d. *Vehicle checks*	X		X	X
e. *Speed limits (type of road/zone)*	X	X	X	
f. *Protective equipment*	X			
g. *Control of drunk driving*	X			
h. *Working conditions of truck drivers*	X			
i. *Driving license (young and old persons)*	X			
j. *Road transit license*				
k. *Certification of transport undertaking*	X			
l. *Restrictions*	X	X	X	
• traffic/parking				
• city centre/zone/road				
• peak hour/day				
• type of vehicle				
m. *Penalties for traffic offences*	X			
2. Public investments				
a. *Roads, streets (design, surface, roadside)*	X	X		
b. *Cycle tracks*	X	X	X	
c. *Roundabouts, squares*	X			
d. *By-passes*	X	X	X	
e. *Freight terminals*				
f. *Intermodal co-ordination*			X	
• park&ride facilities			X	
• combined freight transport	X		X	X
g. *Traffic management system*	X		X	X
h. *Public transport (trams, etc.)*			X	
i. *Emergency services*	X			

Table VI.2. **Instruments available in order to achieve road safety and environmental protection (continued)**

Instruments	Impacts			
	Accident	**Noise**	**Air pollution**	**Energy/CO2**
3. Economic incentives				
a. *Insurance premium*	X			
b. *Vehicle purchase and annual road tax*		X	X	
c. *Fuel tax*				X
d. *Road tolls*		X	X	X
e. *Urban tolls*		X	X	X
f. *Parking charges*		X	X	
g. *Public transport subsidy*			X	
h. *Fines*	X			
4. Communication, management				
a. *Education in schools*	X			
b. *Driver training*	X	X	X	
c. *Information campaigns*	X		X	
d. *Voluntary standards*	X		X	X
e. *Consultation*			X	
f. *Co-ordination between sectors*	X	X	X	X

and safety and the environment. In principle, it is the role of public policy to make decisions on the basis of costs and benefits in order to solve the problems when all the dimensions of conflict have been identified.

The main instruments which have direct consequences on both the environment and road safety are summarised in Table VI.2.

The Scientific Expert Group of the OECD will publish its study in 1997, based on an international survey of actual experience with integrated policies in OECD countries and cities. Without pre-empting the Group's conclusions, it is safe to say that this issue is very complex, and most countries are only at the beginning stage of integration. There are however successful examples of implemented strategies. But much remains to be done.

VI.3. ENHANCING ENVIRONMENTAL TECHNOLOGIES

VI.3.1. Fighting roadnoise

Noise caused by traffic is the nuisance the most often cited by roadside residents. Road noise is growing in OECD countries in particular as a consequence of the sustained expansion of road freight transport. Transport at night is considered a key feature of disquiet. An integrated approach – vehicles, roads, noise screens, buildings – is necessary to arrive at realistic and balanced technical solutions and to streamline the flows of financing and expenditures to attenuate noise effects.

In 1995, a Scientific Expert Group published its work on *Roadside Noise Abatement (6)*. The study presents prevailing regulations and limits in OECD countries as well as the various methods for predicting and measuring noise. The emphasis is on the "road" techniques as part of a dual system : low-noise surfacings and noise barriers, including vegetation and biowalls. Enhanced road layouts – tunnels, cuttings, embankment structures – and planning/land use zoning are also discussed in the study, because these are strategies of noise control that are more global and long-term. Table VI.3 presents the various types of acoustical protection with their associated efficiency and costs.

Notwithstanding – and congruent with OECD's economic principles for environmental policies – a well defined cost/benefit and funding approach is necessary. Taking the German example (and considering only road noise) the total estimated cost would be 750-850 million US$ per year if the aim was to reduce the existing noise by 1 dB(A).

The OECD Group has recommended the following Leq levels to be attained in the ten years to come:

LAeq for existing roads	daytime	65+/-5 dB(A)
	nighttime	55-60 dB(A)
LAeq for new roads	daytime	60+/-5dB(A)
	nighttime	50-55 dB(A)

These figures give targets to achieve. Where there is a will there is a way. But partial results can also be desirable when doing more would be too costly and doing nothing is too damaging.

Table VI.3. **Efficiency and cost of acoustical protections**

Type of acoustical-protection	Average efficiency[1]	Cost per linear metre of road	
		Protection of one side of the road $/m	Protection of both sides of the roads $/m
Noise barrier (with "normal foundations")	6 - 12 dB(A)	600 - 1 600	1 200 - 3 200
"Total " screen	15 - 25 dB(A)	-	6 600
Noise absorbent surfacing with drainage asphalt (o)	3 - 5 dB(A)	-	120 (297) ***
Optimised pavement (porous, semi-thick)	5 - 7 dB(A)	-	less than 1 400
Anti-noise (euphonic) pavement**	5 - 7 dB(A)	-	1 400
Improving soundproofing of facades****	5 - 10 dB(A) *****		
• Collective buildings		3 000	6 000
• Single houses		1 700	3 300
Routes in tunnel	Total protection		
• 2 lanes tunnel		-	10 000-15 000
• 3 lanes tunnel		-	30 000-50 000

[1] Values obtained for single vehicles, 7.50 m from the axis of the rolling lane, 1.20 m from the ground
(o) 10+10 meters paved road
* Protection of top floors
** Especially for low frequencies
*** Considering the bi-annual cost of washing, and the eighth-year cost of recycling, actualised (without taking into account the inflation rate), the figure in parenthesis indicates the pavement cost comparable with the other intervention methods that over the same period of time (15-16 years) have not required any form of intervention (having costs similar to those of installation)
**** Difference between normal window and specialised "antinoise window"
***** Estimated for one house every 30 metres

VI.3.2. Opportunities offered by recycling

As discussed in Chapter IV on Infrastructure, the RTR is building on the 1977 OECD report entitled *Use of Waste Materials and By-Products in Road Construction*. To this end the Programme established a Scientific Expert Group *on Recycling for Road Improvements* in 1995 (8). Its purpose was to assess and report on current practices and policies related to recycling for road construction and make recommendations that would promote recycling of waste and by-product materials.

The new group was formed because the RTR Steering Committee noted the changes in thinking and practice that had taken place in many Member countries during the nearly 20 years that passed between the first report and the creation of the new group. Specifically, by 1995, OECD Member countries had learned that a universal waste management philosophy had to be flexible in that it has to allow for the reassessment of what is classified as waste at any given time. Such a philosophy must adapt as new uses for materials are discovered and be able to accept the possibility that the by-products from one industry can potentially emerge as another industry's usable material.

The report urges road administrations to set priorities for recycling. The first priority is for road officials to find the ways and means to recycle road by-products so that the road construction industry sets an example for others and does not contribute to the adverse environmental impacts associated with the disposal of waste materials. Within the framework of the road industry, therefore, the Group established the following hierarchy of recycling:

1.   Minimise waste production;
2.   Recycle in the parent (road) industry;
3.   Recycle in other (non-road) industries;
4.   Incinerate:
     a. with energy recovery
     b. to reduce volume
5.   Dispose of in a landfill.

In addition to establishing this hierarchy for minimising the environmental impact of road-related construction activities, the report provides advice on the environmental cautions that should be examined prior to using any recycled materials in a road construction or rehabilitation project.

There are various working methods for countries wishing to create or expand their road recycling programmes. An essential element in either process is creating partnerships between the traditional road industry – road administration officials, construction industry, materials suppliers – and relevant environmental agencies and groups. The creation of such partnerships relies upon a clearly delineated set of responsibilities for each party in the partnerships. The report addresses this issue and spells out the appropriate responsibilities necessary for a successful recycling programme.

VI.3.3. *Transport of hazardous materials through road tunnels*

The transport of hazardous materials has been a long standing area of interest in the RTR Programme [e.g. Seminar in Karlstad, (9)]. The focus is now on the *transport of dangerous goods through road tunnels.* The recent accident in the Channel tunnel shows that the topic is crucial and international co-operation is needed to give new directions for future control strategies.

A description of the joint OECD/PIARC project launched in 1995 (10) is given in Section V.6.2. It may be possible that this work will lead to some new assessment of hazardous substances and develop new risk reduction technologies for both transport and tunnel operations. The conclusions and recommendations will be for the attention of all partners involved in the transport of dangerous goods, including national and local decision making bodies, truck operators and truck drivers. Both the EU and ECE, as well as the OECD Environment Directorate are looking forward to receiving the findings of this unique, jointly sponsored research.

VI.3.4. *"Green" Logistics*

Metropolitan areas all over the world and many urban areas in OECD countries are suffering from severe pollution due to road traffic, including freight transport. One method to attenuate these effects is to develop measures that increase the efficiency of freight transport. The use of "city", "green" logistics can provide new opportunities by (11):

- Implementing market forces by the appropriate selection and adoption of environmental policies at national and local levels (see Section II.3).

- Improving the urban freight transport system (joint delivery, increase of the loading rate of urban delivery vehicles by using advanced information system, improvement of parking space for loading and unloading), and

- Developing logistics terminals in suburban areas (concentrating the usage of heavy vehicles on expressways, in order to alleviate congestion in urban areas)

The issue of logistics is dealt with in detail in Chapter II. City logistics concepts are being experimented with in several cities of many OECD countries, but the desired re-organisation of freight transport patterns and new service demands are difficult to reconcile. Cost/benefit considerations and effective public/private partnership are likely to govern future progress.

VI.4. REFERENCES

1. OECD ENVIRONMENT DIRECTORATE. Proceedings of the Conference *Towards Sustainable Transport*, held on 24-27 March 1996, Vancouver (Canada). OECD, Paris.

2. OECD ROAD TRANSPORT RESEARCH (1994). *Environmental Impact Assessment of Roads.* OECD, Paris.

3. Proceedings of the OECD Seminar on *Environmental Impact Assessment of Roads - Strategic and Integrated Approach.* (1994) ANAS, Rome.

4. OECD ROAD TRANSPORT RESEARCH. (1995) *Road Infrastructure Rehabiliation and Safety Strategies in Central and Eastern European Countries.* OECD Document (also available in Russian), Paris.

5. OECD ROAD TRANSPORT RESEARCH (to be published in 1997). *Integrated Environment/Safety Strategies.* OECD, Paris.

6. OECD ROAD TRANSPORT RESEARCH (1995). *Roadside Noise Abatement.* OECD, Paris.

7. Proceedings of the OECD Seminar *on Roadside Noise Abatement.* (1995) CEDEX, Madrid.

8. OECD ROAD TRANSPORT RESEARCH (to be published in 1997). *Recycling for Road Improvement.* OECD, Paris.

9.   OECD. Seminar on *"Strategies For Transporting Dangerous Goods by Road: Safety and Environmental Protection"*. (1992). OECD Environment Monograph no. 66, June 1992. Karlstad, Sweden.

10. OECD ROAD TRANSPORT RESEARCH (to be published in 1998). *Transport of Dangerous Goods through Road Tunnels*. OECD, Paris.

11. OECD ROAD TRANSPORT RESEARCH (1996). *Integrated Advanced Logistics for Freight Transport*. OECD, Paris.

CHAPTER VII: TWO DATABASES

One of the two main fields of activity for the Programme is technology transfer (see also Chapter VIII) and information exchange. The accomplishment of this mission is assured through two major international databases:

- The International Road Research Documentation (IRRD) scheme to assemble and disseminate road related research literature as well as information on research in progress,

- The International Road Traffic and Accident Database (IRTAD) to collect and systematically exchange aggregate accident data from and between OECD Member and other countries.

VII.1. THE INTERNATIONAL ROAD RESEARCH DOCUMENTATION (IRRD) SCHEME

VII.1.1. Mechanism for information exchange and technology transfer

There is an obvious need to appropriately document knowledge and information that is responsive to the changing needs of the professional community. Overlap and double-work can be avoided and innovative technologies can be made accessible on a timely basis by sharing research and experience of different countries. This has been the underlying philosophy of IRRD since its inception in 1965.

Since that time, continuous efforts have been made to improve and enhance the services offered by the IRRD. Over 300,000 scientific references are now electronically available on the IRRD database. The system contains vital and up-to-date bibliographical data and hence provides an indispensable information tool for professionals.

The purpose of the IRRD scheme is to collect, co-ordinate and disseminate all relevant information of interest to practising engineers, managers, researchers, educators and all other actors working in the road and road transport sectors. The references included concern research reports, books, articles from journals and reviews, theses, standards and specifications, conference proceedings, and summaries of research in progress.

VII.1.2. *International co-operative venture*

The IRRD database provides world-wide access to comprehensive bibliographical references on road engineering and road transport. It is open to all countries, institutions and professional users. The computerised database has been in existence since 1972.

Box VII.1. IRRD Membership

Presently, the following OECD countries are active Members of the IRRD scheme:

Australia	Germany	Norway
Austria	Hungary	Spain
Belgium	Ireland	Sweden
Canada	Italy	Switzerland
Denmark	Japan	United Kingdom
Finland	The Netherlands	United States
France		

Brazil, Columbia, the People's Republic of China, and Saudi Arabia are Associate Members and several other Latin American countries are expected to join the scheme in the near future.

Co-operation with the European Conference of Ministers of Transport (ECMT) and the International Railway Union (UIC) is ensured.

All participating countries and institutes co-operate on a continuous basis to maintain and improve the database. Information included is of high quality, and has scientific and technological relevance at the international level. Input to the database is provided by the 31 renowned institutes participating in the co-operative system.

Each participating country is represented by a national IRRD centre and some countries like Austria, France, Germany, Japan and the Netherlands, have also designated additional institutes. The IRRD is a self-financed scheme and the annual subscription fee varies according to the statute of the participating centre: for OECD Member countries, this subscription amounts to 55,000 FF for national IRRD centres and 16,500 FF for additional centres; for IRRD centres outside OECD a special 50% discounted rate (i.e. 27,500 FF) is granted.

The IRRD management is decentralised (1). The OECD Secretariat ensures the administrative and budgetary management of the scheme and reports back to the RTR Steering Committee. The Executive Committee, in existence since 1995, has a supervisory and advisory role with special emphasis on the financial and strategic orientation and future policy direction of the database. The Operational Committee is technical in nature and suggests new measures to improve the database operation. It is assisted by two specialised sub-committees: one dealing with terminology issues and the other with computer and information technologies.

VII.1.3. Multilingual input and database

Informative summaries of published papers and on-going research are included in either English (80 per cent), French (10 per cent) or German (10 per cent) and, since 1994, Spanish.

The database is updated monthly with details of new publications. The progress of on-going research projects is reported at least every three years, and descriptions of new research are added as they become available. 80 per cent of the database references relate to publications and 20 per cent to research projects. Approximately 12,000 references are added each year.

To ensure a high linguistic and quality standard, the IRRD is operated following a rather specific organisation. Co-ordination of data input is ensured by linguistic centres: BASt (Bundesanstalt für Strassenwesen, Bergisch-Gladbach) for German, LCPC (Laboratoire Central des Ponts et Chaussées, Paris) for French, and TRL (Transport Research Laboratory, Crowthorne) for English. With a view to further enhance the usability and international coverage of the database, the need for Spanish language descriptive summaries has been recognised. Since January 1994, certain data (approximately 1,000 currently) are available in Spanish. The Spanish input is undertaken by CEDEX (Centro de Estudios y Experimentacion de Obras Publicas, Madrid). It is expected that

CEDEX will act as the fourth IRRD Co-ordinating Centre for Spanish speaking countries that are likely to join the scheme in the near future.

VII.1.4. Subject areas and fields covered

The primary aim of IRRD Members is to provide information of relevance to all potential IRRD users world-wide. Data collected are not necessarily limited to national coverage. Specialists from IRRD participating institutes select and ensure continuous screening of new publications and research on roads and road transport. Hence, the IRRD scheme offers complete coverage of road and road transport fields (see Box VII.2) and is constantly updated to cover new developments. The previously mentioned Terminology Subcommittee monitors research trends and introduces new descriptors and keywords as needs arise.

Box VII.2. **IRRD Subject Coverage**

- *Traffic and transport*: passenger transport, public transport, freight transport, traffic techniques, traffic management, traffic surveys, social and environmental issues;

- *Road safety -- accidents*: accident statistics, accidents and the road, road safety equipment, road user behaviour, road user education, alcohol and drugs, injuries, fatalities;

- *Vehicles*: Design and safety, comfort, vehicle inspection and control, operation and maintenance, impact on environment;

- *Roads and pavements, bridges, tunnels*: geometric design, layout, drainage, frost, construction methods, equipment, impact on environment, maintenance methods, equipment, winter maintenance;

- *Materials*: bituminous materials, concrete, aggregates, stabilised materials, steel, metals, plastics, etc.

- *Soils and rocks, geotechnics*: soil and rock mechanics, soil and rock properties;

- *Administration and economics*: road and transport statistics, forecasts, financing, programming;

- *Environment: all traffic and road related environmental issues.*

VII.1.5. Access and availability world-wide

Collected information is consolidated monthly and sent to all Member Institutes as well as to the international host server ESA-IRS[3]. Consulting the IRRD database is now easier as the media used are better adapted to respond to specific individuals' requirements. Information can be searched for using a terminal or microcomputer through the ESA-IRS international host computer system (file 43), as well as by consulting the newly developed *TRANSPORT CD* distributed by SilverPlatter.

TRANSPORT CD covers road-related scientific and technical material, transport economics, and the full range of materials and topics found in the U.S. Transportation Research Board (TRB) database. This CD-ROM was produced in response to the availability of evolving information technology and tools as well as to meet the needs expressed by an increasing number of users for improved information access. IRRD Members, jointly with the TRB and the European Conference of Ministers of Transport (ECMT), decided that a CD-ROM would provide an ideal format for this information.

TRANSPORT CD has been available since 1995. It contains some 590,000 publication references from the IRRD, TRIS and TRANSDOC[4] databases. Covering publications back to 1900 and with quarterly updates, *TRANSPORT CD* is a unique tool for finding information in most of the fields covered in the transport sector. At the end of 1996, less than two years after its initiation, 250 institutes from 35 countries were subscribing.

VII.1.6. What are the perspectives?

Information technologies are evolving at a rapid pace. IRRD is therefore now considering new means and venues to further expand information dissemination to new markets and user groups. It is expected that IRRD usage will further develop in non OECD Member countries.

In this context, it should be noted that PIARC[5] has decided to use the IRRD as its official database and to promote its usage amongst its participating members. The World Interchange Network (WIN) scheme – established as a

---

3.
    European Space Agency -- Information Retrieval Service

4.
    Database of the International Co-operation in Transport Economics Documentation (ECMT)

5.
    World Road Association (formerly Permanent International Association of Road Congresses)

follow-up to the OECD sponsored Seminars on Technology Transfer[6] and to the PIARC World Road Congresses held in 1991 and 1995 – will profit from IRRD co-operation.

VII.2. THE INTERNATIONAL ROAD TRAFFIC AND ACCIDENT DATABASE (IRTAD)

The rather good road safety performance exhibited by OECD countries in recent years despite persistent traffic growth can be explained by more than two decades of sustained efforts to improve road safety. In addition, efficient research co-operation between countries in the fields of human behaviour and psychology, vehicle design, road engineering, school education, driver instruction and licensing has contributed greatly to this improved performance. However, road accidents remain an acute problem in OECD Member countries both in terms of human suffering and social and economic costs (see Chapter V). Clearly, road safety assessment will benefit from international comparisons that provide useful benchmarks.

VII.2.1. Scientific approach to advance accident knowledge

Since its inception, the Programme has acknowledged the value of statistical methods in analysing road accidents and, consequently, in better planning road safety policies. The scientific approach was recognised as an essential element to better understand the traffic accident problem and to elaborate effective accident prevention programmes.

Many OECD safety research groups have emphasised the need for enhancing information exchange and expanding international statistics of road accidents with a view to improving the data used for research and policy planning purposes. Indeed, international scientific research in road safety has been limited because national reporting rules and practices differ greatly — not to speak of the variations in underreporting of similar types of accidents — between countries and even within a single country. As a consequence, accident data and frequencies cannot be compared without jeopardising research findings and their interpretation.

---

[6]   Orlando (1990); Seville (1991), Budapest (1992), Helsinki (1996)

VII.2.2. Interactive structure

In 1988, an OECD Group on *Framework for Consistent Traffic and Accident Statistical Data Bases* (2), under the Chairmanship of the Netherlands, investigated the needs for a future joint initiative. 16 countries and the Commission of the European Communities (CEC) participated. This was the starting point of the International Road Traffic and Accident Database (IRTAD) that was created in the early 1990s.

Three purposes were in mind when initiating IRTAD: strategic — synergetic — organic. *Strategic* because no traffic safety policy can be conducted in an ad hoc approach. *Synergetic* because scientific international co-operation is essential in traffic safety where markets are global. *Organic* because a workable framework and a solid organisation provide the basis for successful joint efforts.

Today, 18 countries have designated national co-ordinating centres to contribute to the IRTAD work. The countries are shown in Box VII.3.

Box VII.3. **IRTAD National Representation**		
Australia	France	Norway
Austria	Germany	Spain
Canada	Hungary	Sweden
Czech Republic	Japan	Switzerland
Denmark	The Netherlands	United Kingdom
Finland	New Zealand	United States

Because a wide range of public and private groups have a vested interest in the struggle for enhanced road safety, the IRTAD is structured in such a way that the private sector — the automobile industry such as Ford, Mercedes Benz, Volkswagen and the Japan Automobile Research Institute, the insurance industry like Verband der Schadensversicherer e.V. and safety associations such as the AA Foundation — can have an active role in IRTAD and benefit from the experience of national research bodies. Altogether 15 associate institutions are subscribing to IRTAD. Because one of its aims is to be internationally representative on a global scale, IRTAD is open to all non-OECD member countries.

VII.2.3. Unique set of internationally comparable data

The aggregate national data from 27 countries are collected on a regular basis and are made comparable by internationally agreed upon factors based on an accurate review of the national differences. The IRTAD system is based on 240 variables in such a way that aggregate national data of fatal accidents and some exposure measures can be combined in a meaningful way to compare fatal risks. As to parameters, responsiveness and comparability, it goes beyond the statistical tabulations of the Economic Commission for Europe, the International Road Federation or the ECMT which provide analyses of selected accident trends. Complementing these international database efforts, IRTAD statistics can be used for a variety of national purposes by the participating agencies.

The main topics covered by IRTAD are illustrated in Box VII.4. Road traffic and accident data are available on a yearly basis from 1970 onwards with 1965 as a reference year.

Box VII.4. Principal Sets of Road Traffic and Accident Data

- Population figures with a breakdown by age group.
- Vehicle population with a breakdown by vehicle types.
- Kilometrage classified by network areas and vehicle types.
- Number of injury accidents classified by road network areas.
- Fatality figures with a breakdown by types of road usage, age groups and network areas.
- Hospitalised with a breakdown by types of road usage, ages groups and network areas.
- Network length classified by network areas.
- Modal split.
- Area of state.
- Risk values: fatalities, hospitalised and injury accidents related to population or kilometrage figures.
- Monthly accident data for selected countries (three key variables).

IRTAD offers a necessary framework for international evaluations of accident data. It is hosted by the German Federal Highway Research Institute (BASt). Membership is based on a yearly subscription fee.

The 33 participating institutions have a common data format and definitions when inputting and harmonising the data. IRTAD Members and BASt monitor the quality, consistency, plausibility and compliance of the data. To overcome data and statistical problems they exchange information and expertise taking stock of the experience of high level road safety statisticians, professionals and researchers. Special reports on selected accident research topics – such as under-reporting, traffic exposure, definitions, etc. – have been published.

IRTAD data are electronically available through a combined searching strategy. Users can obtain a copy of the database at regular intervals for installation on their own micro-computer and have free on-line access to the database on the BASt mainframe via remote data transmission for the most recent figures. User-friendliness is an IRTAD slogan. To this end, ready-to-use retrieval and instruction courses are made available to IRTAD Members, and software is continuously improved and developed. Also, information on the general use of the database is provided on a regular basis.

VII.2.4. Effective co-operation through networking

The IRTAD is a traffic accident analysis tool that stimulates international discussion on harmonisation of definitions and spurs improvements in data collection and comparison. It is a living database in that new concepts are considered instrumental because they can be used as an analytical tool and a statistical foundation for road safety policy formulation.

Workshops for policy and research users are organised to consolidate ongoing work, disseminate the most recent statistical accident research and assist in the decision-making process. The latest Seminar on *International Road Traffic and Accident Databases* (3) held in Helsinki in September 1995 reflects this dynamism. It was attended by approximately 100 participants from OECD, Member countries, Central and Eastern European Countries and international organisations such as the European Commission and the European Conference of Ministers of Transport.

Pursuing improvements is at the core of the IRTAD. Further extension to OECD non-member countries and stronger liaison with other international organisations are essential. This database can help to identify conceptual requirements necessary for the development of integrated accident counter-measures and to provide road authorities and policy makers with a greater awareness of what is at stake world-wide. Co-operation with the EU's proposed

CARE (Community database with dissagregated data on injury Accidents on the Road in Europe) data scheme is therefore a key priority for the future.

Participation in IRTAD means long term improvement of accident statistics and international comparability in concert with the needs of OECD Member and other countries. Equally importantly, participation in IRTAD can be an incentive for more highly focused and effective safety measures and actions.

VII.3. REFERENCES

1. OECD ROAD TRANSPORT RESEARCH (1995). International Road Research Documentation: Manual of Practice. OECD, Paris.

2. SWOV (1988) for the OECD ROAD TRANSPORT RESEARCH PROGRAMME. Framework for Consistent Traffic and Accident Statistical Databases. SWOV, Leidschendam.

3. OECD ROAD TRANSPORT RESEARCH (1995). International Road Traffic and Accident Databases: Helsinki Seminar Proceedings. OECD, Paris.

CHAPTER VIII: TECHNOLOGY TRANSFER

VIII.1. PROMOTING T^2 INITIATIVES

For much of its existence, the Programme has been a forum for technology transfer (T^2) and knowledge exchange. From the outset the goal was to provide a mechanism for technology co-operation.   A prime priority was the International Road Research Documentation (IRRD) scheme established in 1965 to assure the systematic and formalised exchange of scientific and technical information (see Chapter VII).

In the recent past, the Programme's activities were significantly enhanced to develop dialogue and co-operation with non-Member countries. In response to the emergence of significant economic players in Asia and Latin America, as well as the opening up of Central and East Europe, the activities undertaken with countries of these regions encompass a range of areas including research and analysis, policy dialogue and advice, workshops, seminars, expert groups, and limited technical assistance.

The non-member outreach activities have focused primarily on two types of economies, those that can be characterised as economies in transition and developing countries. This characterisation has proven useful in the past as it was perceived that the needs of these two distinctive groups were different. The former group of countries – i.e. mainly CEECs and NIS – can be described as undergoing a major shift from a relatively inflexible rail-based surface transport system to one dominated by more flexible road systems that provide higher quality service. The latter category of countries primarily aim at improving the quality and extent of road systems in order to serve their economic and social development plans. In addition to recognising the unique needs of different types of countries, the OECD/RTR has also sought to be in the forefront of finding useful and practical ways to identify and address the key generic issues where appropriate technologies and improved management processes will bring progress and benefits.

The Programme's recent T^2 initiatives can be summarised as follows:

T^2 generic policy seminars

- *Industrialised Nations*, Orlando, USA, 11-14 November 1990;
- *Developing Nations and Economies in Transition*, Seville, Spain, 17-20 September 1991;
- *Road Technology Transfer and Diffusion for Central and Eastern European Countries*, Budapest, Hungary, 12-14 October 1992.
- *Road and Traffic Technology Transfer*, (under OECD patronage), Helsinki, 30 May-1 June 1996

T^2 priority projects

- *Transport Planning and Road Design in Developing Countries* – two strategic studies (1975-80);
- *Road Maintenance* – three manuals (1980-95);
- *Road Safety* – three regional conferences (Addis Ababa, Kuala Lumpur and Sao Paolo), (1989-95);
- *Bridge Engineering* – Jakarta conference in September 1996 under the WRA/PIARC umbrella (1).

T^2 programme for CEECs and NIS

- *The Budapest Conference* – see above (1992);
- *The Scientific Expert Group on Road Strengthening in Central and Eastern European Countries.*(1992);
- *Six Workshops on Road Maintenance* (1993-1994);
- *Six Workshops on Road Safety* (1994-1995);
- *Two Workshops on Policy Matters* (1994-1995);
- *Concluding Workshop Conference* – Ljubljana, Slovenia, October 1995.

VIII.2. EXPLORING OPTIONS IN CEECs and NIS

In order to ensure that the newly emerging market economies in the Central and East European Countries (CEEC) and the Newly Independent States (NIS) would not be penalised in their efforts to compete in international markets, the

Programme undertook a variety of initiatives to bolster their road programmes. Part of these activities were financed through grants from EU/PHARE, OECD Member countries and industry.

VIII.2.1. Technology Transfer Seminar

Because technology transfer can be a cost-effective means to significantly address basic educational, technical and organisational structural problems, the Steering Committee placed emphasis on reviewing adjustment policies and processes. With this guidance, the Programme organised the *Seminar on Road Technology and Diffusion for Central and East European Countries* in 1992 in Budapest (2). The seminar emphasised the need to maintain and enhance the scientific and technological capacities and networks of CEECs.

In summary, the seminar achieved the following:

- Enhanced technology transfer and information exchange by way of the two RTR databases;
- Assessed road-related T^2 and technical assistance initiatives and projects and adopted a co-ordinated approach for them (see Box VIII.1);
- Prepared the foundation for future bi- and multilateral technical exchange and co-operation;
- Established the groundwork for a network of experts and professionals in key road sectors.

It was recognised that T^2 remains a complex process. Donors must ensure that the key knowledge essential to technology adoption or adaptation is transferred. User needs must be defined clearly and then the most appropriate technology, strategy and operation can be identified.

In order to be successful, firstly a conceptual understanding is required. This means a rational, structured process involving professionals, consultancy, demonstration projects and focused trials, supported by public/private partnerships in the client/recipient group or country. Secondly, commitment is essential at all levels of the user organisation. The effort should be proactive, selecting the right people to work with the novel technology or method. Reform-minded managers should be sought. Thirdly, to bring about a change in work culture, T^2 must reach the various groups and echelons. In order to reach grass roots and assure true adaptation, facilitators and targeted training are necessary.

T^2 friendly regulations need to be introduced. To create responsiveness by the professional community and the agencies, the target should be results-oriented – not necessarily aiming at the use of high technology, but the "right", "appropriate" technology. This is well known, but still a barrier in many countries of the world. Horizontal co-operation with agencies in other sectors so as to converge with national goals is desirable and allowance must be made for sufficiently long periods of transition.

Box VIII.1. T^2 Toolbox	
DATABASES	• IRRD: International Road ResearchDocumentation • IRTAD: International Road Traffic and Accident Database
TRAINING	• Initial training • Post-graduate training • In-service training • Seminars, workshops • Professional training • Study tours • Tutored learning • Movies, videos
RESEARCH - EXPERT GROUPS	• Joint research programmes • Exchange of researchers • Laboratory equipment
RULES	• Regulations • Standardisation • Specifications • Manuals, guidelines
TECHNICAL ASSISTANCE	• Aid programmes
JOINT PROJECTS	• Co-operative projects • Twinning • Partnerships
SETTING OF INSTITUTIONAL FRAMEWORK	• Legal, regulatory and administrative texts • Organisation of agencies • Management methods • Working methods • Computer software • Expert services
INDUSTRY LEAD/ INVOLVEMENT	• Joint ventures • Transfer of patents, licences and know-how • Supply of equipment

The Budapest Seminar also proposed to launch the World Technology Transfer Network, asking WRA and the international donor agencies to take the lead in launching this initiative. This was a follow-up to the 1991 OECD Seville Seminar conclusions presented to the XIXth World Congress of WRA in Marrakech calling for a global T^2 entity and clearinghouse. The independent World Interchange Network (WIN) was established at the September 1995 World Road Congress in Montreal. It allows those in need of technical advice about appropriate and applicable technologies to identify a choice of experts/institutions and/or technology providers and manufacturers. The OECD sponsored Seminar on Road and Traffic Technology,held in June 1996 in Helsinki, explored further actions to extend the use of the Network (see also Section VIII.4).

VIII.2.2. Key issue: Infrastructure upkeep

Among the first direct co-operative activities instigated by the OECD was the creation in 1992 of an Expert Group for *Road Strengthening: Short-term Strategies and Techniques for CEECs*. This was a joint OECD/CCET activity in which representatives from Bulgaria, the Czech Republic, Hungary, Lithuania, Poland, Romania, Russia and the Slovak Republic participated, as well as experts from Austria, Denmark, Finland, France, Germany, Greece, Italy, the Netherlands, Spain, Switzerland, Turkey, the United States, the World Bank and the IRF. In 1993 the Group culminated their work with the publication of a report entitled *Road Strengthening in Central and Eastern European Countries* (3).

The report addressed several issues of importance to the CEECs including: the current situation as regards finance and budget needs; technical issues; rating systems and assessment of pavement conditions; diagnosis and design for strengthening pavements; strengthening measures and techniques; rapid strengthening measures; and the management and implementation of rehabilitation operations.

The report also made specific recommendations for future co-operation to improve the road and transport situation in the CEECs (see Box VIII.2). According to a World Bank estimate, between US $ 10 to 15 billion (thousand million) in rehabilitation investments are necessary (1992 figures).

Of special importance in the near and distant future is the spreading of the use of the 11.5 tonne European (EC) standard axle load which will introduce increased traffic stresses in the existing pavements that have not been designed for such loads. In other words, the fourth power law[7] and the special protection needed for under–designed pavements during the spring thaw period are two crucial elements that should drive CEECs – as well as donor and aid agencies – to press on with proper maintenance and strengthening strategies.

Box VIII.2. Technical needs and highway priorities in CEECs

Management, Maintenance and Rehabilitation

- Long-term pavement behaviour
- Pavement evaluation and condition parameters
- Road data collection and data management
- Pavement management systems
- Use of local materials in road works
- Drainage systems for pavements
- Use of new technologies for pavement rehabilitation
- Better technologies for road markings
- Bridge management, inspection and rehabilitation

Road Safety and Heavy Freight Traffic

- Short- and medium-term assessment of the impact of increasing transport by road on the major routes of national networks
- Impacts of the 11.5 t axle load onthe road network
- Targeted safety initiatives for road infrastructure
- Driver education and training, especially heavy truck drivers

Financial, Economic and Social Issues

- Prioritisation and allocation of road funds
- The motorway network development and the relationship between East and West
- Assistance in contracting procedures, joint venture initiatives and project preparation facilities for financing institutions
- Development and operation of toll roads
- T^2 regional centres and information dissemination
- Training and education of staff in the road sector
- Road works and related environmental effects

---

[7] The relevant effect, in terms of pavement damage, of the 11.5 t compared to the 10 t axle load is $(\dfrac{11.5}{10})^4 = 1.75$, i.e; the 11.5 t axle has a 75 per cent higher damage potential.

VIII.2.3. Policy advice and technical aid through programme of Workshops

One of the most important outcomes of the Budapest seminar was a mandate for the Programme to arrange for a series of 14 workshops and a concluding conference (see Box VIII.3). The purpose of the workshops was to share state-of-the-art and practice information to road managers, engineers, planners, and others to facilitate and ease the implementation of extensive infrastructure and rehabilitation programmes and advise safety authorities on accident countermeasures. Additionally, the workshops were geared to provide the basis for increased personal contacts and the exchange of viewpoints on the need for long-term and co-ordinated initiatives.

Box VIII.3. OECD Workshops for CEECs and NIS

- *Bituminous materials for road construction and maintenance*
  (7-10 December 1993, Budapest, Hungary)
- *Road rehabilitation and strengthening*
  (25-29 April 1994, Arad, Romania)
- *Accident data system*
  (16-20 May 1994, Jurmala, Latvia)
- *Road aggregates*
  (25-28 May 1994, Sofia, Bulgaria)
- *Management of existing bridges*
  (12-16 September 1994, Kaunas, Lithuania)
- *Road maintenance management*
  (26-30 September 1994, Warsaw, Poland)
- *Education and training of drivers*
  (3-6 October 1994, Warsaw, Poland)
- *Road winter maintenance*
  (18-21 October 1994, Prague, Czech Republic)
- *Infrastructure design and road safety*
  (15-18 November 1994, Prague, Czech Republic)
- *Roads and road transport in Black Sea countries*
  (23-25 November 1994, Istanbul, Turkey)
- *Vehicle inspection*
  (13-16 December 1994, Budapest, Hungary)
- *Environmental impact evaluation of roads*
  (6-10 February 1995, Prague, Czech Republic)
- *Automobile insurance and traffic safety*
  (10-12 May 1995, Tallinn, Estonia)
- *Children's safety/education*
  (9-10 October 1995, Warsaw, Poland)

The broad success of these initiatives was the fact that they were responsive to the expressed needs of the road and road transport authorities in the targeted regions. They provided focused technical assistance and technology transfer to the CEECs and their designated road and road transport experts. In addition, the workshops helped to promote the overall goals of the Programme and its efforts to disseminate the collective expertise of the Member countries to the CEECs and NIS. The workshops gave the CEECs and NIS the opportunity to rapidly advance their road expertise and practices in the face of daunting and explosive growth in the demand for higher quality and more extensive road systems and effective traffic safety programmes.

In order to bring together the findings and experience from these Workshops, a *Concluding Conference of the CEEC and NIS Series of Workshops* (4) was held on 19-20 October 1995 in Ljubljana, Slovenia to do the following:

1. Evaluate the results and effectiveness of the series of workshops;
2. Identify subjects for future workshops (if necessary);
3. Consider other technology transfer mechanisms to support the CEEC and NIS;
4. Clarify the specific needs and requirements of the CEEC and NIS.

The last of these items was especially important to the OECD and the RTR Steering Committee. It was recognised that it would be impossible for any further co-operative assistance initiative, whether bilateral or multilateral, to succeed unless the goals of such assistance were tailored to meet the real or perceived needs of the receiving country. The conference results revealed a very positive response from the CEEC and NIS concerning the workshop series. It was felt that the workshops contributed to raising awareness of some issues, elevated technical practice in some areas, and brought about policy and organisational changes that would prove beneficial to the CEEC and NIS road and traffic safety administrations. Though there were some specific ideas for additional workshops in a few select areas, there was general agreement that there was not a need for a new, large series of workshops.

VIII.2.4. Laying out the future road map

To develop a carefully calibrated policy for the future, assessments were solicited from the CEECs and NIS in a follow-up enquiry in 1995/96 on what their critical and specific needs were. The answers, coupled with discussions at the conference, brought to light the fact that, in reference to other technology transfer mechanisms, there was a need for some major Seminars on broad, horizontal themes such as quality assurance, environmental planning and road management policy, especially in regard to the European 11.5 tonne axle load and freight traffic in general, as well as targeted road safety projects.

It became apparent that follow-up activities could be quite different for different countries based on the national or local context. Also, there were diverging interests amongst the OECD countries and consultants involved. It was agreed that bilateral and small multilateral projects could be an answer if recipient countries had long term work plans in place. This is an area of continuing interest for the OECD/RTR and, as such, an Advisory Panel was established in late 1996 to begin identifying and working on the next set of activities to address these needs.

VIII.3. CONTRIBUTING TO SUSTAINABLE DEVELOPMENT

In the same manner that the Programme reaches out to support the European economies in transition, so has it sought to develop programmes and activities that address the specialised needs of developing regions of the world. Clearly, the involvement of a multilateral institution with long–term expertise, such as RTR, can be an important factor in assisting developing regions who have the most to gain from transport investment.

The World Bank has reported (5) that the economic rate of return on World Bank-supported highway projects amounts to 29 per cent. This is the highest rate of return in comparison to any other type of infrastructure investment. The Bank has also documented a clear connection between economic development and the use of roads. As a country moves from Low-income to Middle-income to High-income, road transport becomes the dominant means of transport with increasing problems of traffic safety and, more recently, environmental nuisance.

VIII.3.1. Emphasis on road maintenance

One of the problems that plagues developing countries is the maintenance of existing roads. Though maintenance problems themselves are quite similar in developing and industrialised countries, several prominent features separate them. First, developing countries borrow a substantial portion of their maintenance funds from international aid organizations. Secondly, the road networks in developing countries are generally not fully developed which creates competition between the application of funds for new road construction as opposed to maintenance of existing roads. Thirdly, in developing countries the major network of primary and secondary roads is usually complemented by the existence of extensive feeder roads and tracks. Finally, according to the World Bank, only approximately one-third of major roads in developing countries are in good condition. Another one-third are in fair condition and require attention. The last one-third of roads are in poor condition requiring reconstruction. These situations point to the essential role that organisation and management can play to ensure that maintenance policies are effective and durable.

To advance on these issues, an OECD Scientific Expert Group that included World Bank experts undertook a project directed at *Road Monitoring for Maintenance in Developing Countries* (6). Their work culminated in 1990 with an OECD inspection manual and a damage catalogue for developing countries. As a follow-up, the RTR Steering Committee created a new group charged with the following two objectives:

1.   Validate the operational procedures for visual assessment and routine inspections which were proposed in the manual; and

2.   Define the main structure of a road maintenance management system.

Their work resulted in a 1995 OECD publication entitled *Road Maintenance Management Systems in Developing Countries* (7) (available in English, French and Spanish). The report emphasises the need to know the condition of the road network in order to satisfy three purposes: to assess maintenance needs; to search for an optimum policy; and to monitor the policy implemented and its effects on the road network.

The focus is on Road Management Systems (RMS) whose aim is to find the optimum strategy which provides the greatest benefit for the community while taking into account the constraints related to available resources. By using models it is possible to examine the incidence of each maintenance scenario in terms of changing road conditions and transport cost. Three types of data are necessary for optimisation: the relationship between the level of maintenance and vehicle operating costs; damage models and the results of maintenance work; and the relationship between pavement distress and performance. State-of-the-art developments in OECD countries and in regard to the World Bank's HDM model are reviewed in the report.

The scheduling of periodic maintenance is a critical part of an RMS and consists of selecting which sections are to receive which treatment and in which order of priority, in compliance with the objectives of the strategy which has been laid down beforehand. Scheduling for the road networks in developing countries is more difficult, as — due to maintenance backlog — rehabilitation and maintenance exist side by side. Maintenance scheduling requires information about the network, technical tools and skills, an understanding of the context and an effective institutional structure.

In all this, a flexibly structured data bank is the core of the RMS. The report describes how an organisation can design a data bank, input data, create a reference system, access information and maintain the integrity of the data. It illustrates how the data bank provides the basis to examine various applications of maintenance techniques, levels and costs to fully optimise the scheduling of maintenance works. The data bank will serve to perform analysis using the technico-economic assessment model chosen in order to compare several long-term strategies and define annual and multi-annual scheduling of works. Because a maintenance management system is a dynamic process, data for monitoring is critical to help in assuring the continuity of the RMS and compliance with the financial and institutional commitments, as well as technical requirements, i.e. pavement performance and level of service.

A final item that received great attention from the Group was training. Clearly, staff training is a key problem which must be tackled to ensure the successful implementation of a maintenance management system. The report contains the building blocks of a training programme which takes account of both the context of developing countries and the conclusions of the OECD report on *Monitoring of Roads for Maintenance Management in Developing Countries*. Four areas of training are identified for the "management", "general co-ordination", "field engineer" and "road inspector" levels.

VIII.3.2. Scaling down road accidents

One of the areas that is critical in the development of road systems throughout the world is safety. World-wide there are more than 500 000 fatalities per year and about 15 million people are injured in road accidents. The associated social and economic costs are immeasurable. For developing countries, the problem is far worse. For example, between 1968 and 1985, traffic deaths in Asia and Africa grew by an average 235 per cent. Whereas, in the developed world, traffic deaths fell by 25 per cent over the same period. As another measure of the problem, traffic deaths as measured against car ownership reveals that the developing regions of the world suffer the highest death rates on their roads (8).

Understanding the relationship between technical inefficiency and road safety problems, the OECD/RTR has been on the forefront in trying to stem the tide in road deaths. Through a series of regional conferences/technology transfer initiatives, the OECD has sought to bring recognised safety experts together with traffic authorities from the non-Member countries suffering most from safety problems. The 1995, 1996 and early 1997 conferences for Latin America and the Caribbean, Asia, and Africa built on previous successful efforts in Africa and Asia. Though individual conferences were designed for the varying situations of the regions and countries concerned, the general themes of the conferences were to:

1. Raise awareness of the problem;

2. Stimulate national and regional road safety plans and programmes, including Research & Development;

3. Share information on successful practical approaches to improve road safety;

4. Bring about reduced numbers of road traffic fatalities.

The *Road Safety Conference for Latin America and the Caribbean* was held 4-7 December 1995 in Sao Paulo, Brazil (9). In addition to the OECD, the conference was co-sponsored by the Panamerican Institute of Highways (PIH) and the Governments of Brazil, the State of Sao Paulo, Spain and Portugal. The organisers established objectives for the conference along the general themes listed above. The conference was attended by 400 persons from 30 countries. As a result of discussions during the conference, *"The Sao Paulo Declaration"* was drafted and adopted. The document declared that traffic accidents leading to deaths and injuries incur economic costs equal to 1 to 3 percent of the countries's Gross Domestic Products. One of the key reasons for drafting the Declaration was to ensure that this message was delivered to senior Government officials throughout the region.

Box VIII.4. Measures and commitments agreed in accordance with the Declaration

The conference participants agreed to the following measures or commitments as spelled out in the Declaration:

1. Seek support from Government's for National Traffic Safety Programmes that cover all aspects of highway safety;

2. Establish a Latin America and Caribbean data bank aligned with other at the international level, especially the OECD/RTR International Road Traffic Accident Database, in order to facilitate traffic safety knowledge, studies, standards, policy development, and research.

3. Promote the establishment of a National Traffic Safety Committee in each of the participating countries;

4. Institutionalise, through the PIH, a catalogue of courses for the specific field of traffic safety;

5. Encourage technology exchange and co-operation among the Pan American countries in road safety;

6. Implement and update the adoption of the Inter-American Handbook of Traffic Control devices;

7. Promote the elaboration of standards for the safety inspection of vehicles and drivers;

8. Promote national and local traffic safety events to establish awareness of the problem;

9. Communicate the intentions of the Declaration and the conference results to the organises of the Third African Road Safety Conference;

10. Promote a Second Highway Safety Conference for Latin America and the Caribbean in 1998.

The purpose of the conference is not to put the OECD/RTR Programme in the lead of implementing their outcomes. Rather, the purpose is to empower individual countries to recognise the problem and to take independent but hopefully harmonised action to improve road safety in their countries. In this regard many of the countries are acting on the conference results. Thus, the last recommendation within the Declaration calling for a second conference is designed to keep interest alive in and the focus on road safety issues in these countries. At a second conference, countries will be asked to report on how they have made concrete steps to improve their road safety situation. This will serve two purposes. First, it will allow countries to share their experiences and success stories. Secondly, it will help to institutionalise the concept of road safety by continuing to keep it within the programme agendas of the countries in the region.

In the same spirit of the above conference, the *Second Conference on Asian Road Safety: Evaluation and Priorities* (2nd CARS) was organised by an international committee jointly chaired by China and Japan and held from 28-31 October 1996 in Beijing, People's Republic of China (10). Again, about 400 participants from 30 countries attended, as well as several international organisations including the World Bank, the Asian Development Bank and the Road Engineering Association of Asia and Australasia. The objectives of the 2nd CARS included first and foremost to follow-up on the results of the 1st CARS held in October 1993 in Kuala Lumpur (11). In addition, the conference was intended to continue bringing attention to Asian road safety problems by continuing to monitor changes in the safety situation. The focus was on keeping road safety officials in the Asian countries informed and aware of the latest developments in road safety technologies and practices, emphasising their potential effectiveness.

The main output of the 2nd CARS was a document entitled *"The Beijing Agenda"*. The Beijing Agenda, like the Declaration described above, documented specific recommendations to serve as a guide for all Asian Nations to solve their road safety problems. The key areas and recommendations suggested are summarised in Box VIII.5.

As with the Latin American and Caribbean region, the Asian nations felt that the subject was of paramount importance for their region. Thus, in order to maintain interest and encourage continuing implementation of the recommendations outlined above, the countries agreed to hold a third conference in Seoul, South Korea in 1999.

Building on two earlier conferences, the *Third African Road Safety Conference* is planned to take place in Pretoria, South Africa on 14-17 April 1997 (12) as a joint initiative between the United Nations Economic Commission for Africa, OECD and South Africa. As with the 2nd CARS, the objectives of this conference will be to continue reporting on the experiences and successes of implementing road safety programmes in African nations. The conference will also focus on continuing the momentum for developing and implementing integrated approaches for road safety. Finally, the conference is designed to assist, through technology and information exchange, the African nations to further develop strategies and actions plans for road safety programmes in their nations.

Box VIII.5. **The Beijing Agenda**

1. *Road Safety Policy*: Road safety policy should be an integrated part of road transport policy which links infrastructure development with economic development.

2. *Organisation*: It was strongly recommended that an appropriate institutional and policy framework be established that would ensure an effective organisational and managerial structure.

3. *Financing, Costs and Benefits*: Because economic losses due to road accidents in the Asia-Pacific region now amount to $US 20 billion, investments in traffic safety are justified. Therefore, sustainable sources of public funding for road safety should be developed.

4. *Road Safety Programmes*: Road safety programme development begins with an inventory of major accident problems and then are most effective when associated with detailed targets. There is a need for immediate, short-term programmes that ensure positive impacts in a relatively short period of time.

5. *Road Safety and Training*: Whole life educational programmes as well as educational and training programmes for professional drivers should be developed.

6. *Road Safety Evaluation*: Effective Systems for accident data collection should be instituted. As well, monitoring and evaluation of road safety measures should be carried out by independent auditors and the costs and effects of road safety actions should be properly assessed.

7. *Road Safety Measures and Priorities*: Safety of vulnerable road users is crucial. This is especially true for motorcycle safety as it is a prevalent means of personal transportation in Asia. Additionally, traffic calming and speed reductions should be considered for specific situations in Asia. Finally, low-cost engineering measures should be emphasised by highway authorities throughout the region.

8. *Exchange of Knowledge and Experience*: Technology transfer and exchange of experience in road safety should be enhanced through the establishment of an appropriate regional and international network. As well, regular meetings on road safety should be held.

VIII.4. SHARING KNOWLEDGE AND EXPERTISE GLOBALLY

In addition to focusing on regional issues and problems, the OECD/RTR Programme engages in global efforts to improve the technical expertise of road officials and the efficiency of road programmes. As pointed out above, one notable area where the Programme has been instrumental in the creation of the World Interchange Network (WIN or INTERCHANGE).

WIN is based on the concept of global information and knowledge sharing. It has its origins in the concerns and recommendations expressed during the three OECD Technology Transfer Seminars of 1990, 1991 and 1992. A Founders' Conference was held in June 1994 in Casablanca, under the auspices of the World Road Association (WRA) where the following main aims of the INTERCHANGE network were identified: improve access to world-wide road expertise; accelerate technology dissemination and implementation; facilitate assessments of domestic versus international practices; increase complementarity of Research and Development efforts; strengthen existing organisations and encourage creation of new ones where needed.

As designed, WIN operates on the concept of a network of nodes having geographic importance or specialised expertise. The affiliation principle restrains expenses primarily to the node or user level rather than at the site of a single organisation, thus relieving countries (or organisations) of the need for central international financing. Through this network, question/answer contacts and links are facilitated. Likewise, WIN creates the unique opportunity to put a list of specialised technical experts at the disposal of those people and countries who may want to take advantage of their services. All in all, WIN creates a dynamic and new dimension to global knowledge sharing based on state of the art communications technology. The INTERCHANGE Secretariat is currently housed in Quebec, Canada with the Provincial Government.

The Programme has contributed to the development of WIN. RTR was represented at the Founders' Conference and participated in the drafting of the INTERCHANGE "Declaration of Intent". The OECD has also invited, on regular occasions, the participation of the World Road Association and the WIN Board of Directors to Steering Committee Meetings to brief the members on the WIN concept and developments. Likewise, the OECD has undertaken several initiatives such as IRRD and IRTAD promotion in non-Member countries and the identification of OECD experts available to non-Member countries for advice and consultation. The OECD sees WIN as being at the fore in exploring new avenues and mechanisms for technology transfer and knowledge exchange.

In this regard, the OECD will continue to support the development and expansion of WIN during the next Triennial Programme.

VIII.5. PLANNING THE FUTURE OF NON-MEMBER OUTREACH

Judging from the experience gained, the T^2 initiatives of the Programme were very well received. Professional contacts and links were established with experts and officials from over 80 non-Member countries. All new OECD Members – Mexico, Hungary, the Czech Republic, Poland and Korea – have found it useful to join and support RTR as a special Budget II programme of the OECD.

So far, the Programme has found the separation between the European transitional economies and the developing regions of the world useful. However, recognising that these two groupings do not necessarily include all those that could be characterised either as dynamic non-Members or emerging market economies, the Programme has identified the need to develop an integrated approach to the ever-increasing and diverse demands of all non-Member countries. For this reason, and in keeping with the overall thrust of the OECD to expand activities with non-Member countries, the Programme has reoriented its efforts under the umbrella category of "Outreach Activities". This strategic approach allows the Programme to better consolidate its activities as well as maximise the benefits derived from investments in co-operative ventures with non-Member countries.

As a first step to initiating a comprehensive, but selective, outreach strategy, the Programme established in 1996 an Advisory Panel on Outreach Activities. The overall purpose of the Panel is to advise the Steering Committee on outreach activities that cover all facets of the RTR Programme and to ensure that the proposed actions are responsive to the stated scientific and technical needs of non-Member countries. The benefits derived from outreach activities are to be maximised for the limited funds available for these activities. Clearly, the Programme has to avoid over-stretching of resources. The Advisory Panel will therefore seek the active involvement of and co-ordination with other international organisations. During the next Triennial Programme, the Panel is specifically directed to:

- Establish and maintain a list of recognised OECD/RTR experts available for road policy and technical support requested by non-Member countries;
- Advise the RTR Secretariat on how to respond to requests for such support from non-Member countries;
- Propose to the Steering Committee targeted outreach activities within the confines of the earmarked RTR Programme resources;
- Assign members of organising groups for outreach activities approved by the Steering Committee. Oversee planning and preparations for outreach activities;
- Co-ordinate OECD involvement in outreach activities with those of other international bodies;
- Promote the International Road Research Database and the International Traffic Accident Database in co-operation with the established RTR Committees;
- Evaluate the outcome of OECD-supported outreach activities and report the outcome to the Steering Committee along with recommendations as appropriate;
- Disseminate information on the conclusions and recommendations as well as follow-up actions taken.

Through the combined efforts of the Advisory Panel, the Steering Committee, and the RTR Secretariat, the outreach efforts of the Programme should continue to bring about positive results for both OECD and non–Member countries. A case in point is the TRILOG initiative (see Chapter II), involving APEC, North America and the EC. This reflects the crucial importance of the global market place and the deepening and widening of the interdependence of world regions. Trade without transport is impossible. The TRILOG project is therefore designed to encompass common world-wide concerns and centres on advanced logistics as a guiding force in global integration and exchange. In this vein, it is envisioned that, in a number of traditional and innovative road sectors, the next three year effort will aid in promoting open and purposeful interaction between the transport officials of OECD Member countries and their counterparts in non-Member countries.

VIII.6. REFERENCES

1. MINISTRY OF PUBLIC WORKS OF INDONESIA/PIARC (1996). *International Seminar on Bridge Engineering and Management in Asian Countries.* Jakarta, Indonesia. September 1996.

2. OECD ROAD TRANSPORT RESEARCH PROGRAMME (1992). *Seminar on Road Technology and Diffusion for Central and East European Countries.* Budapest, Hungary. 12-14 October 1992.

3. OECD ROAD TRANSPORT RESEARCH PROGRAMME (1993). *Road Strengthening in Central and Eastern Europe.* OECD, Paris.

4. OECD ROAD TRANSPORT RESEARCH (1995). *Concluding Conference of the CEEC and NIS Series of Workshops.* Ljubljana, Slovenia. 19-20 October 1995.

5. THE WORLD BANK (1994). *World Development Report 1994.* The World Bank, Washington, D.C.

6. OECD ROAD TRANSPORT RESEARCH (1990). *Road Monitoring for Maintenance in Developing Countries.* OECD, Paris .

7. OECD ROAD TRANSPORT RESEARCH (1995). *Road Maintenance Management Systems in Developing Countries.* OECD, Paris.

8. TRANSPORT RESEARCH LABORATORY (1991). *Towards Safer Roads in Developing Countries.* TRL, Berkshire, England.

9. PAN AMERICAN INSTITUTE OF HIGHWAYS/OECD (1995). *Road Safety Conference for Latin America and the Caribbean.* Sao Paulo, Brazil. 4-7 December 1995.

10. SOCIETY OF TRAFFIC ENGINEERING, CHINA/OECD (1996). *Second Conference on Asian Road Safety: Evaluation and Priorities.* Beijing, People's Republic of China. 28-31 October 1996.

11. MALAYSIAN GOVERNMENT/OECD (1993). *Proceedings of the Conference on Asian Road Safety 1993.* Kuala Lumpur.

12.   ECA/OECD (1997). *Third African Road Safety Conference.* Pretoria, South Africa. 14-17 April 1997.

ANNEX A

Table A.1. **Scientific Expert Groups**

TITLE	CHAIRMAN	PUBLICATION
Transport of Dangerous Goods through Road Tunnels	D. Lacroix (France)	1998/99
DIVINE	P. Sweatman (Australia)	1997
Performance Indicators for the Road Sector	A. Talvitie (The World Bank)	1997
Recycling for Road Improvements	C. Nemmers (United States)	1997
Safety of Vulnerable Road Users	N. Muhlrad (France)	1997
Safety Theory, Models & Research Methodologies	K. Rumar (Sweden)	1997
Integrated Environment/Safety Strategies	A.HH Jansson (Finland) & S. Lassarre (France)	1997
Training of Truck Drivers	Rapporteur: J.L. Tardif (Canada)	1996
Integrated Advanced Logistics for Freight Transport	M. Manheim (United States)	1996
Repairing Bridge Substructures	S. Gordon (United States)	1995
Roadside Noise Abatement	G. Camomilla (Italy)	1995
Road Maintenance Management Systems in Developing Countries	P. Autret (France)	1995
Improving Road Safety by Attitude Modification	H.J. Johansen (Denmark)	1994
Congestion Control and Demand Management	S. Strickland and S Lockwood (United States)	1994

Targeted Road Safety Programmes	F. Wegman (Netherland)	1994
Environmental Impact Assessment of Roads	C. Lamure (France)	1994
Road Maintenance and Rehabilitation: Funding and Allocation Strategies	A. Talvitie (The World Bank)	1994
Road Strengthening in Central and Eastern European Countries	P. Sulten (Germany)	1993
Marketing of Traffic Safety	R.D. Huguenin (Switzerland)	1993
Intelligent Vehicle Highway Systems: Review of Field Trials	K. Rumar (Sweden)	1992
Advanced Logistics and Road Freight Transport	H. Kawashima (Japan)	1992
Dynamic Loading of Pavements	P. Sweatman (Australia)	1992
Bridge Management	M. El-Marasy (Netherlands)	1992
Cargo Routes: Truck Roads and Networks	J.P. Boender (Netherlands)	1992

Table A.2. **Seminars/Workshops/Conferences/Symposia**

TITLE	Committee Chairman	Place	Date
Intermodal Networks and Logistics	D. Diaz Diaz (Mexico)	Mexico City, Mexico	June 1997
Third African Road Safety Conference	M. Bongoy (ECA) B. Horn (OECD)	Pretoria, South Africa	April 1997
Symposium on "Globalization & Advanced Logistics"	Y. Okano (Japan)	Fukuoka, Japan	November 1996
Second Conference on Asian Road Safety	T. Nakaoka (Japan) Y. Pang (China)	Beijing, China	October 1996
Evaluation and Development of Road Transport Research Programmes	J.P. Medevielle (France)	Annecy, France	October 1996
Prioritisation of multimodal transport infrastructure	G.J.D. Offerman (Netherlands)	Netherlands	May 1996
Transport of Dangerous Goods through Road Tunnels – Risk Assessment and Decision-making Process: Methodologies, Models, Tools	Co-ordinator: N.O. Jorgensen (Denmark)	Oslo, Norway	March 1996
Road Safety Conference for Latin America and the Caribbean	T.H. Monteiro Penteado & H.A. Moreira (Brazil)	Sao Paulo, Brazil	December 1995
Roadside Noise Abatement	CEDEX (Spain)	Madrid, Spain	November 1995
Concluding Conference of the CEEC and NIS Series of Workshops	J. Canny (United States)	Ljubljana, Slovenia	October 1995
International Road Traffic and Accident Databases	M.J. Koornstra (Netherlands)	Helsinki, Finland	September 1995
Environmental Impact Assessment of Roads - Strategic and Integrated Approach	C. Lamure (France)	Palermo, Italy	June 1994
Seminar on Advanced Road Transport Technologies	H. Yamada (Japan)	Omiya, Japan	June 1994
OECD Seminar on "Strategic Planning for Road Research Programs"	R. Betsold (United States)	Williamsburg, United States	October 1993

Conference on Asian Road Safety 1993 (CARS '93)	M. Koshi (Japan) Dato Jamilus Hussein (Malaysia)	Kuala Lumpur, Malaysia	October 1993
Workshop on Congestion Control	S. Strickland (United States)	Barcelona, Spain	March 1993
Seminar on Road Technology and Diffusion for Central and East European Countries	A. Lacleta (Spain)	Budapest, Hungary	October 1992
2nd OECD Workshop on Knowledge-based Expert Systems in Transportation	J.A. Wentworth (United States)	Montreal, Canada	June 1992
Strategies For Transporting Dangerous Goods by Road: Safety and Environmental Protection	U. Bjurman (Sweden)	Karlstad, Sweden	June 1992

Please note that a series of 14 Workshops for CEECs and NIS were held in 1993-1995.

ANNEX B

STEERING COMMITTEE MEMBERS IN 1996-97

AUSTRALIA	Mr Peter MAKEHAM
	Mr Ian JOHNSTON
AUSTRIA	Mr Heinz LUKASCHEK
BELGIUM	Mr C. van den MEERSSCHAUT
	Mr Jean Claude HOUTMEYERS
	Mr Roger CAIGNIE
CANADA	Mrs Nicole PAGEOT
	Mr Derek SWEET
	Mr Christopher HEDGES
CZECH REPUBLIC	Mr Petr POSPISIL
	Mr Zdenek TRCKA
DENMARK	Mr Niels HELBERG
	Mr Jorgen CHRISTENSEN (**Chairman**)
FINLAND	Mr Jukka ISOTALO
	Mr Asko SAARELA
	Mr M. SALUSJÄRVI
FRANCE	Mr Jean-François COSTE
	Mr Pierre CHANTEREAU (**Vice-Chairman**)
GERMANY	Mr Hermann DEFFKE
	Mr Karl-Heinz LENZ

GREECE	Ms Maria SAKKI
HUNGARY	Mr Sandor TOMBOR Mr Boldizsar VASARHELYI
ICELAND	Mr Helgi HALLGRIMSSON
IRELAND	N.N.
ITALY	Mr Pasquale CIALDINI Mr Enrico SAMMARTINO
JAPAN	Mr Kunihiko TAKADA (**Vice-Chairman**) Mr Masakazu NAKAGAWA Mr Seizo TSUJI Mr Nobuaki SATO Mr Masayoshi DOBASHI Mr Motohide YOSHIKAWA
LUXEMBOURG	Mr Nico MARMANN
MEXICO	Mr Daniel DIAZ DIAZ Mr Oscar de BUEN RICHKARDAY
NETHERLANDS	Mr Peter M.W. ELSENAAR (**Vice-Chairman**) Mr Matthijs J. KOORNSTRA
NEW ZEALAND	Mr Nigel MOUAT
NORWAY	Mr Knut OSTMOE Mr Kaare FLAATE
POLAND	N.N.
PORTUGAL	Ms Maria DA PIEDADE CORREA ROBERTO
SOUTH KOREA	N.N.
SPAIN	Mr Oscar ALVAREZ ROBLES Mr César LOZANO (**Vice-Chairman**)

SWEDEN	Mr Thomas KORSFELDT
SWITZERLAND	Mr Konrad RIEDER
	Mr Bernard PERISSET
TURKEY	Mr Ali GÜRGÜR
	Mr Asim CAVUSOGLU
UNITED KINGDOM	Mr David METZ
	Ms Finella McKENZIE
UNITED STATES	Mr Joseph CANNY (**Vice-Chairman**)
	Mr Robert J. BETSOLD
EEC	Mr Wim A.G. BLONK
OECD/DSTI	Mr Risaburo NEZU
	Mr Thomas ANDERSSON
OECD/RTR	Mr Burkhard HORN
	Mr Claude MORIN
	Mr Patrick HASSON
	Mr Ceallach LEVINS
	Mr Takeshi YOSHIDA
	Ms Marie-Dominique GORRIGAN
	Ms Véronique FEYPELL
	Mrs Carmen DUSSIEUX
	Ms Marie Christine du BOUËTIEZ
	Ms Rozanna HERRING

The report Outlook 2000 was compiled by the RTR Secretariat.

MAIN SALES OUTLETS OF OECD PUBLICATIONS
PRINCIPAUX POINTS DE VENTE DES PUBLICATIONS DE L'OCDE

AUSTRALIA – AUSTRALIE
D.A. Information Services
648 Whitehorse Road, P.O.B 163
Mitcham, Victoria 3132   Tel. (03) 9210.7777
Fax: (03) 9210.7788

AUSTRIA – AUTRICHE
Gerold & Co.
Graben 31
Wien I   Tel. (0222) 533.50.14
Fax: (0222) 512.47.31.29

BELGIUM – BELGIQUE
Jean De Lannoy
Avenue du Roi, Koningslaan 202
B-1060 Bruxelles Tel. (02) 538.51.69/538.08.41
Fax: (02) 538.08.41

CANADA
Renouf Publishing Company Ltd.
5369 Canotek Road
Unit 1
Ottawa, Ont. K1J 9J3   Tel. (613) 745.2665
Fax: (613) 745.7660

Stores:
71 1/2 Sparks Street
Ottawa, Ont. K1P 5R1   Tel. (613) 238.8985
Fax: (613) 238.6041

12 Adelaide Street West
Toronto, QN M5H 1L6   Tel. (416) 363.3171
Fax: (416) 363.5963

Les Éditions La Liberté Inc.
3020 Chemin Sainte-Foy
Sainte-Foy, PQ G1X 3V6 Tel. (418) 658.3763
Fax: (418) 658.3763

Federal Publications Inc.
165 University Avenue, Suite 701
Toronto, ON M5H 3B8   Tel. (416) 860.1611
Fax: (416) 860.1608

Les Publications Fédérales
1185 Université
Montréal, QC H3B 3A7   Tel. (514) 954.1633
Fax: (514) 954.1635

CHINA – CHINE
Book Dept., China Natinal Publiations
Import and Export Corporation (CNPIEC)
16 Gongti E. Road, Chaoyang District
Beijing 100020 Tel. (10) 6506-6688 Ext. 8402
(10) 6506-3101

CHINESE TAIPEI – TAIPEI CHINOIS
Good Faith Worldwide Int'l. Co. Ltd.
9th Floor, No. 118, Sec. 2
Chung Hsiao E. Road
Taipei   Tel. (02) 391.7396/391.7397
Fax: (02) 394.9176

CZECH REPUBLIC –
RÉPUBLIQUE TCHÈQUE
National Information Centre
NIS – prodejna
Konviktská 5
Praha 1 – 113 57   Tel. (02) 24.23.09.07
Fax: (02) 24.22.94.33
E-mail: nkposp@dec.niz.cz
Internet: http://www.nis.cz

DENMARK – DANEMARK
Munksgaard Book and Subscription Service
35, Nørre Søgade, P.O. Box 2148
DK-1016 København K   Tel. (33) 12.85.70
Fax: (33) 12.93.87

J. H. Schultz Information A/S,
Herstedvang 12,
DK – 2620 Albertslung   Tel. 43 63 23 00
Fax: 43 63 19 69
Internet: s-info@inet.uni-c.dk

EGYPT – ÉGYPTE
The Middle East Observer
41 Sherif Street
Cairo   Tel. (2) 392.6919
Fax: (2) 360.6804

FINLAND – FINLANDE
Akateeminen Kirjakauppa
Keskuskatu 1, P.O. Box 128
00100 Helsinki

Subscription Services/Agence d'abonnements :
P.O. Box 23
00100 Helsinki   Tel. (358) 9.121.4403
Fax: (358) 9.121.4450

*FRANCE
OECD/OCDE
Mail Orders/Commandes par correspondance :
2, rue André-Pascal
75775 Paris Cedex 16 Tel. 33 (0)1.45.24.82.00
Fax: 33 (0)1.49.10.42.76
Telex: 640048 OCDE
Internet: Compte.PUBSINQ@oecd.org

Orders via Minitel, France only/
Commandes par Minitel, France
exclusivement : 36 15 OCDE

OECD Bookshop/Librairie de l'OCDE :
33, rue Octave-Feuillet
75016 Paris   Tel. 33 (0)1.45.24.81.81
33 (0)1.45.24.81.67

Dawson
B.P. 40
91121 Palaiseau Cedex   Tel. 01.89.10.47.00
Fax: 01.64.54.83.26

Documentation Française
29, quai Voltaire
75007 Paris   Tel. 01.40.15.70.00

Economica
49, rue Héricart
75015 Paris   Tel. 01.45.78.12.92
Fax: 01.45.75.05.67

Gibert Jeune (Droit-Économie)
6, place Saint-Michel
75006 Paris   Tel. 01.43.25.91.19

Librairie du Commerce International
10, avenue d'Iéna
75016 Paris   Tel. 01.40.73.34.60

Librairie Dunod
Université Paris-Dauphine
Place du Maréchal-de-Lattre-de-Tassigny
75016 Paris   Tel. 01.44.05.40.13

Librairie Lavoisier
11, rue Lavoisier
75008 Paris   Tel. 01.42.65.39.95

Librairie des Sciences Politiques
30, rue Saint-Guillaume
75007 Paris   Tel. 01.45.48.36.02

P.U.F.
49, boulevard Saint-Michel
75005 Paris   Tel. 01.43.25.83.40

Librairie de l'Université
12a, rue Nazareth
13100 Aix-en-Provence   Tel. 04.42.26.18.08

Documentation Française
165, rue Garibaldi
69003 Lyon   Tel. 04.78.63.32.23

Librairie Decitre
29, place Bellecour
69002 Lyon   Tel. 04.72.40.54.54

Librairie Sauramps
Le Triangle
34967 Montpellier Cedex 2 Tel. 04.67.58.85.15
Fax: 04.67.58.27.36

A la Sorbonne Actual
23, rue de l'Hôtel-des-Postes
06000 Nice   Tel. 04.93.13.77.75
Fax: 04.93.80.75.69

GERMANY – ALLEMAGNE
OECD Bonn Centre
August-Bebel-Allee 6
D-53175 Bonn   Tel. (0228) 959.120
Fax: (0228) 959.12.17

GREECE – GRÈCE
Librairie Kauffmann
Stadiou 28
10564 Athens   Tel. (01) 32.55.321
Fax: (01) 32.30.320

HONG-KONG
Swindon Book Co. Ltd.
Astoria Bldg. 3F
34 Ashley Road, Tsimshatsui
Kowloon, Hong Kong   Tel. 2376.2062
Fax: 2376.0685

HUNGARY – HONGRIE
Euro Info Service
Margitsziget, Európa Ház
1138 Budapest   Tel. (1) 111.60.61
Fax: (1) 302.50.35
E-mail: euroinfo@mail.matav.hu
Internet: http://www.euroinfo.hu//index.html

ICELAND – ISLANDE
Mál og Menning
Laugavegi 18, Pósthólf 392
121 Reykjavik   Tel. (1) 552.4240
Fax: (1) 562.3523

INDIA – INDE
Oxford Book and Stationery Co.
Scindia House
New Delhi 110001   Tel. (11) 331.5896/5308
Fax: (11) 332.2639
E-mail: oxford.publ@axcess.net.in

17 Park Street
Calcutta 700016   Tel. 240832

INDONESIA – INDONÉSIE
Pdii-Lipi
P.O. Box 4298
Jakarta 12042   Tel. (21) 573.34.67
Fax: (21) 573.34.67

IRELAND – IRLANDE
Government Supplies Agency
Publications Section
4/5 Harcourt Road
Dublin 2   Tel. 661.31.11
Fax: 475.27.60

ISRAEL – ISRAËL
Praedicta
5 Shatner Street
P.O. Box 34030
Jerusalem 91430   Tel. (2) 652.84.90/1/2
Fax: (2) 652.84.93

R.O.Y. International
P.O. Box 13056
Tel Aviv 61130   Tel. (3) 546 1423
Fax: (3) 546 1442
E-mail: royil@netvision.net.il

Palestinian Authority/Middle East:
INDEX Information Services
P.O.B. 19502
Jerusalem   Tel. (2) 627.16.34
Fax: (2) 627.12.19

ITALY – ITALIE
Libreria Commissionaria Sansoni
Via Duca di Calabria, 1/1
50125 Firenze   Tel. (055) 64.54.15
Fax: (055) 64.12.57
E-mail: licosa@ftbcc.it

Via Bartolini 29
20155 Milano   Tel. (02) 36.50.83

Editrice e Libreria Herder
Piazza Montecitorio 120
00186 Roma   Tel. 679.46.28
Fax: 678.47.51

Libreria Hoepli
Via Hoepli 5
20121 Milano   Tel. (02) 86.54.46
                      Fax: (02) 805.28.86

Libreria Scientifica
Dott. Lucio de Biasio 'Aeiou'
Via Coronelli, 6
20146 Milano   Tel. (02) 48.95.45.52
                      Fax: (02) 48.95.45.48

JAPAN – JAPON
OECD Tokyo Centre
Landic Akasaka Building
2-3-4 Akasaka, Minato-ku
Tokyo 107   Tel. (81.3) 3586.2016
                      Fax: (81.3) 3584.7929

KOREA – CORÉE
Kyobo Book Centre Co. Ltd.
P.O. Box 1658, Kwang Hwa Moon
Seoul   Tel. 730.78.91
                      Fax: 735.00.30

MALAYSIA – MALAISIE
University of Malaya Bookshop
University of Malaya
P.O. Box 1127, Jalan Pantai Baru
59700 Kuala Lumpur
Malaysia   Tel. 756.5000/756.5425
                      Fax: 756.3246

MEXICO – MEXIQUE
OECD Mexico Centre
Edificio INFOTEC
Av. San Fernando no. 37
Col. Toriello Guerra
Tlalpan C.P. 14050
Mexico D.F.   Tel. (525) 528.10.38
                      Fax: (525) 606.13.07
E-mail: ocde@rtn.net.mx

NETHERLANDS – PAYS-BAS
SDU Uitgeverij Plantijnstraat
Externe Fondsen
Postbus 20014
2500 EA's-Gravenhage Tel. (070) 37.89.880
Voor bestellingen: Fax: (070) 34.75.778

Subscription Agency/Agence d'abonnements :
SWETS & ZEITLINGER BV
Heereweg 347B
P.O. Box 830
2160 SZ Lisse   Tel. 252.435.111
                      Fax: 252.415.888

NEW ZEALAND –
NOUVELLE-ZÉLANDE
GPLegislation Services
P.O. Box 12418
Thorndon, Wellington Tel. (04) 496.5655
                      Fax: (04) 496.5698

NORWAY – NORVÈGE
NIC INFO A/S
Ostensjoveien 18
P.O. Box 6512 Etterstad
0606 Oslo   Tel. (22) 97.45.00
                      Fax: (22) 97.45.45

PAKISTAN
Mirza Book Agency
65 Shahrah Quaid-E-Azam
Lahore 54000   Tel. (42) 735.36.01
                      Fax: (42) 576.37.14

PHILIPPINE – PHILIPPINES
International Booksource Center Inc.
Rm 179/920 Cityland 10 Condo Tower 2
HV dela Costa Ext cor Valero St.
Makati Metro Manila   Tel. (632) 817 9676
                      Fax: (632) 817 1741

POLAND – POLOGNE
Ars Polona
00-950 Warszawa
Krakowskie Prezdmiescie 7 Tel. (22) 264760
                      Fax: (22) 265334

PORTUGAL
Livraria Portugal
Rua do Carmo 70-74
Apart. 2681
1200 Lisboa   Tel. (01) 347.49.82/5
                      Fax: (01) 347.02.64

SINGAPORE – SINGAPOUR
Ashgate Publishing
Asia Pacific Pte. Ltd
Golden Wheel Building, 04-03
41, Kallang Pudding Road
Singapore 349316   Tel. 741.5166
                      Fax: 742.9356

SPAIN – ESPAGNE
Mundi-Prensa Libros S.A.
Castelló 37, Apartado 1223
Madrid 28001   Tel. (91) 431.33.99
                      Fax: (91) 575.39.98
E-mail: mundiprensa@tsai.es
Internet: http://www.mundiprensa.es

Mundi-Prensa Barcelona
Consell de Cent No. 391
08009 – Barcelona   Tel. (93) 488.34.92
                      Fax: (93) 487.76.59

Libreria de la Generalitat
Palau Moja
Rambla dels Estudis, 118
08002 – Barcelona
          (Suscripciones) Tel. (93) 318.80.12
          (Publicaciones) Tel. (93) 302.67.23
                      Fax: (93) 412.18.54

SRI LANKA
Centre for Policy Research
c/o Colombo Agencies Ltd.
No. 300-304, Galle Road
Colombo 3   Tel. (1) 574240, 573551-2
                      Fax: (1) 575394, 510711

SWEDEN – SUÈDE
CE Fritzes AB
S–106 47 Stockholm   Tel. (08) 690.90.90
                      Fax: (08) 20.50.21

For electronic publications only/
Publications électroniques seulement
STATISTICS SWEDEN
Informationsservice
S-115 81 Stockholm   Tel. 8 783 5066
                      Fax: 8 783 4045

Subscription Agency/Agence d'abonnements :
Wennergren-Williams Info AB
P.O. Box 1305
171 25 Solna   Tel. (08) 705.97.50
                      Fax: (08) 27.00.71

Liber distribution
Internatinal organizations
Fagerstagatan 21
S-163 52 Spanga

SWITZERLAND – SUISSE
Maditec S.A. (Books and Periodicals/Livres
et périodiques)
Chemin des Palettes 4
Case postale 266
1020 Renens VD 1   Tel. (021) 635.08.65
                      Fax: (021) 635.07.80

Librairie Payot S.A.
4, place Pépinet
CP 3212
1002 Lausanne   Tel. (021) 320.25.11
                      Fax: (021) 320.25.14

Librairie Unilivres
6, rue de Candolle
1205 Genève   Tel. (022) 320.26.23
                      Fax: (022) 329.73.18

Subscription Agency/Agence d'abonnements :
Dynapresse Marketing S.A.
38, avenue Vibert
1227 Carouge   Tel. (022) 308.08.70
                      Fax: (022) 308.07.99

See also – Voir aussi :
OECD Bonn Centre
August-Bebel-Allee 6
D-53175 Bonn (Germany) Tel. (0228) 959.120
                      Fax: (0228) 959.12.17

THAILAND – THAÏLANDE
Suksit Siam Co. Ltd.
113, 115 Fuang Nakhon Rd.
Opp. Wat Rajbopith
Bangkok 10200   Tel. (662) 225.9531/2
                      Fax: (662) 222.5188

TRINIDAD & TOBAGO, CARIBBEAN
TRINITÉ-ET-TOBAGO, CARAÏBES
Systematics Studies Limited
9 Watts Street
Curepe
Trinidad & Tobago, W.I. Tel. (1809) 645.3475
                      Fax: (1809) 662.5654
E-mail: tobe@trinidad.net

TUNISIA – TUNISIE
Grande Librairie Spécialisée
Fendri Ali
Avenue Haffouz Imm El-Intilaka
Bloc B 1 Sfax 3000   Tel. (216-4) 296 855
                      Fax: (216-4) 298.270

TURKEY – TURQUIE
Kültür Yayinlari Is-Türk Ltd.
Atatürk Bulvari No. 191/Kat 13
06684 Kavaklidere/Ankara
                      Tel. (312) 428.11.40 Ext. 2458
                      Fax : (312) 417.24.90

Dolmabahce Cad. No. 29
Besiktas/Istanbul   Tel. (212) 260 7188

UNITED KINGDOM – ROYAUME-UNI
The Stationery Office Ltd.
Postal orders only:
P.O. Box 276, London SW8 5DT
Gen. enquiries   Tel. (171) 873 0011
                      Fax: (171) 873 8463

The Stationery Office Ltd.
Postal orders only:
49 High Holborn, London WC1V 6HB
Branches at: Belfast, Birmingham, Bristol,
Edinburgh, Manchester

UNITED STATES – ÉTATS-UNIS
OECD Washington Center
2001 L Street N.W., Suite 650
Washington, D.C. 20036-4922
                      Tel. (202) 785.6323
                      Fax: (202) 785.0350
Internet: washcont@oecd.org

Subscriptions to OECD periodicals may also
be placed through main subscription agencies.

Les abonnements aux publications périodiques
de l'OCDE peuvent être souscrits auprès des
principales agences d'abonnement.

Orders and inquiries from countries where Dis-
tributors have not yet been appointed should be
sent to: OECD Publications, 2, rue André-Pascal, 75775 Paris Cedex 16, France.

Les commandes provenant de pays où l'OCDE
n'a pas encore désigné de distributeur peuvent
être adressées aux Éditions de l'OCDE, 2, rue
André-Pascal, 75775 Paris Cedex 16, France.

                                        12-1996

OECD PUBLICATIONS, 2, rue André-Pascal, 75775 PARIS CEDEX 16
PRINTED IN FRANCE
(77 97 02 1 P) ISBN 92-64-15477-9 – No. 49389 1997